Interactive
Mathematics Program®

INTEGRATED HIGH SCHOOL MATHEMATICS

Small World, Isn't It?

FIRST EDITION AUTHORS:
Dan Fendel, Diane Resek, Lynne Alper, and Sherry Fraser

CONTRIBUTORS TO THE SECOND EDITION:
Sherry Fraser, IMP for the 21st Century
Jean Klanica, IMP for the 21st Century
Brian Lawler, California State University San Marcos
Eric Robinson, Ithaca College, NY
Lew Romagnano, Metropolitan State College of Denver, CO
Rick Marks, Sonoma State University, CA
Dan Brutlag, Meaningful Mathematics
Alan Olds, Colorado Writing Project
Mike Bryant, Santa Maria High School, CA
Jeri P. Philbrick, Oxnard High School, CA
Lori Green, Lincoln High School, CA
Matt Bremer, Berkeley High School, CA
Margaret DeArmond, Kern High School District, CA

Key Curriculum Press

Second Edition   I M P

This material is based upon work supported by the National Science Foundation under award numbers ESI-9255262, ESI-0137805, and ESI-0627821. Any opinions, findings, and conclusions or recommendations expressed in this publication are those of the authors and do not necessarily reflect the views of the National Science Foundation.

Key Curriculum Press
1150 65th Street
Emeryville, California 94608
email: editorial@keypress.com
www.keypress.com
10 9 8 7 6 5 4 3 2 1   14 13 12 11
ISBN 978-1-60440-049-6
Printed in the United States
of America

Project Editors
Mali Apple, Josephine Noah, Sharon Taylor

Project Administrators
Emily Reed, Juliana Tringali

Professional Reviewers
Rick Marks, Sonoma State University, CA
D. Michael Bryant, Santa Maria High School, CA, retired

Accuracy Checker
Carrie Gongaware

First Edition Teacher Reviewers
Daniel R. Bennett, Moloka'i High School, HI
Maureen Burkhart, Northridge Academy High School, CA
Dwight Fuller, Ponderosa High School, CA
Daniel S. Johnson, Silver Creek High School, CA
Brian Lawler, California State University San Marcos, CA
Brent McClain, Vernonia School District, OR
Susan Miller, St. Francis of Assisi Parish School, PA
Amy C. Roszak, Cottage Grove High School, OR
Carmen C. Rubino, Silver Creek High School, CA
Barbara Schallau, East Side Union High School District, CA
Kathleen H. Spivack, Wilbur Cross High School, CT
Wendy Tokumine, Farrington High School, HI

First Edition Multicultural Reviewers
Genevieve Lau, Ph.D., Skyline College, CA
Arthur Ramirez, Ph.D., Sonoma State University, CA
Marilyn Strutchens, Ph.D., Auburn University, AL

Copyeditor
Brandy Vickers

Interior Designer
Marilyn Perry

Production Editor
Andrew Jones

Production Director
Christine Osborne

Editorial Production Supervisor
Kristin Ferraioli

Compositor
Lapiz Digital Services, Kristin Ferraioli

Art Editor/Photo Researcher
Maya Melenchuk

Technical Artists
Lapiz Digital Services, Laurel Technical Services, Maya Melenchuk

Illustrators
Taylor Bruce, Deborah Drummond, Tom Fowler, Briana Miller, Evangelia Philippidis, Sara Swan, Diane Varner, Martha Weston, April Goodman Willy

Cover Designer
Jeff Williams

Printer
Lightning Source, Inc.

Mathematics Product Manager
Elizabeth DeCarli

Executive Editor
Josephine Noah

Publisher
Steven Rasmussen

CONTENTS

Small World, Isn't It?—Slope, Derivatives, and Exponential Growth

Small World, Isn't It?

Slope, Derivatives, and Exponential Growth

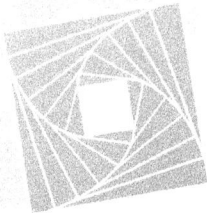

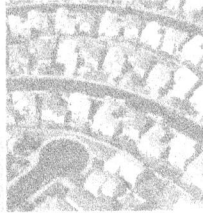

Small World, Isn't It?—Slope, Derivatives, and Exponential Growth

As the World Grows

How many people are there in the world today? What about a century or two ago? What about in the future?

The central problem of this unit involves analyzing world population trends. You will begin by studying population data since 1650, with the goal of determining how long it will take until people are "squashed up against one another."

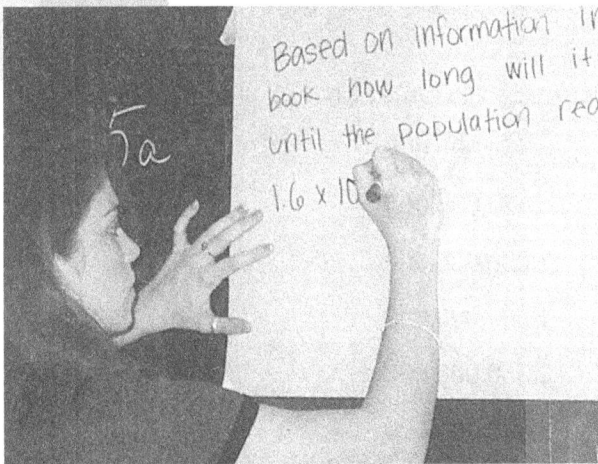

Lisa Newton rephrases the unit's central problem in her own words.

A Crowded Place

Everyone knows the world's population is increasing. The table shows the estimated world population over the past several centuries.

Suppose this pattern of data continues. How long do you think it will take until we are all squashed up against one another?

You may find these facts useful.

- The total surface area of the earth is approximately 197,000,000 square miles.

- Approximately 29.2% of the earth's total surface area is land.

As you consider this question, make note of any difficulties you encounter or any issues you think need clarification.

Year	Estimated Population
1650	470,000,000
1750	629,000,000
1850	1,030,000,000
1900	1,550,000,000
1950	2,560,000,000
1960	3,040,000,000
1970	3,710,000,000
1980	4,450,000,000
1990	5,280,000,000
2000	6,090,000,000
2005	6,480,000,000

Source: U.S. Census Bureau.

The More, the Merrier?

Isn't life more interesting when there are lots of people around? Then why should people worry about a population explosion? Your task in this POW is to study some aspect of our world's rapid population growth and to find out why some people are concerned.

Write a report on your findings, which may include your own opinions and conjectures about why things are the way they are. Graphs, drawings, and tables may help you communicate your ideas.

Choose from these topics, or think of your own topic for your report.

- Population growth and the food supply
- Pros and cons of limiting family size
- Comparison of population growth in different countries
- Population growth and the environment
- Ways to cope with increasing population
- Traffic congestion

How Many of Us Can Fit?

The activity *A Crowded Place* asks you when we will all get "squashed up against one another." That probably won't ever happen, but you may be curious about how many people that would require.

1. How many square miles of land surface area are there on the earth? Use the information in *A Crowded Place* to answer this question.

2. How many square feet of land area are there on the earth? Use the fact that 1 mile is equal to 5280 feet, so 1 square mile is 5280^2 square feet, which is 27,878,400 square feet.

3. In 2005, the world population was about 6.48 billion people. If this population were spread evenly over the earth's land area, about how many square feet of area would each person have, on average?

4. Suppose we interpret the phrase "squashed up against one another" to mean that each person has about 1 square foot of area to call his or her own. How many people would there need to be for us to really be squashed up against one another?

How Many More People?

Part I: The Graph

Making a graph can help you get some ideas about how the world population has grown over the centuries.

1. Choose an appropriate scale and plot the data from *A Crowded Place*. On your graph, show all the data points given in that activity. Label the data points with their coordinates.

Part II: Average Increases

2. a. Find the increase in population between the years 1650 and 1900.

   b. Explain how the increase you found shows up on the graph.

   c. Find the average increase in population *per year* between 1650 and 1900.

   d. What would the portion of the graph from 1650 to 1900 look like if the population had increased by the same number of people each year of this time period?

3. a. Find the increase in population between 1900 and 1950.

   b. Explain how the increase you found shows up on the graph.

   c. Find the average increase in population per year between 1900 and 1950.

   d. What would the portion of the graph from 1900 to 1950 look like if the population had increased by the same number of people each year of this time period?

Part III: Making Comparisons

4. Which interval—1650 to 1900, or 1900 to 1950—has the greater average increase in population per year? How could you answer this question simply by looking at your graph?

Growing Up

The graph shows the average height for boys ages 0 to 6 in the United States around the middle of the twentieth century.

1. Suppose the average amount of growth shown during the first year after birth were to continue during the second year. What would be the average height for boys on their second birthdays?

2. The section of the graph for ages 3 to 6 is nearly straight. What does that mean in terms of the average growth for boys during those years?

3. a. How much does the average height for boys increase between ages 3 and 4?

   b. Express your answer to part a as a percentage of the average height for boys at age 3.

4. a. How much does the average height for boys increase between ages 5 and 6?

   b. Express your answer to part a as a percentage of the average height for boys at age 5.

   c. Compare your answers for Questions 4a and 4b to your answers for Questions 3a and 3b.

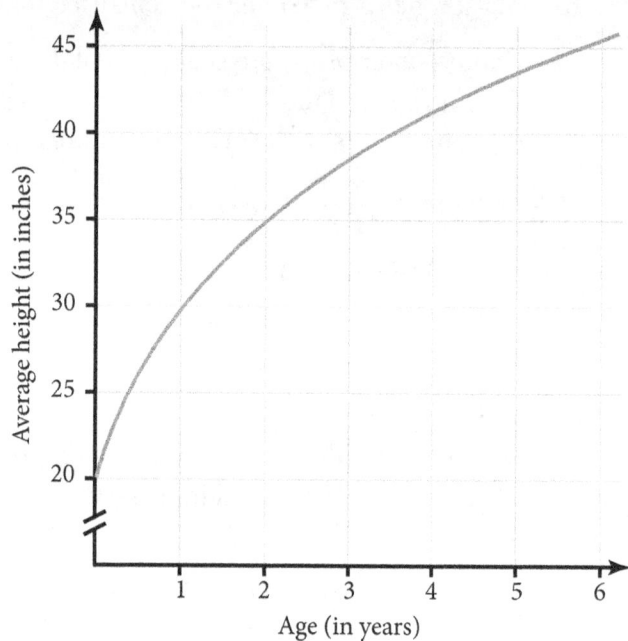

Data source: Department of Pediatrics, State University of Iowa, 1943.

Average Growth

Most growth—such as the growth of living things, the economy, or organizations—is uneven. Population growth is no exception. Over the next few days, you'll look at some examples of growth and rates of change. You will focus your attention on how you can use graphs to understand what's going on.

When a rate of growth is changing, it sometimes helps to examine the *average* rate of growth. When a growth rate is constant, the graph of the situation is a straight line. The slope of that line is the standard numeric method for measuring the steepness of the graph. In these activities, you'll learn a new way to think about slope.

Leilani Juan and Liana DeGracia begin their investigation of growth and rates of change.

Story Sketches

1. This graph describes a wagon train's progress along the Overland Trail.

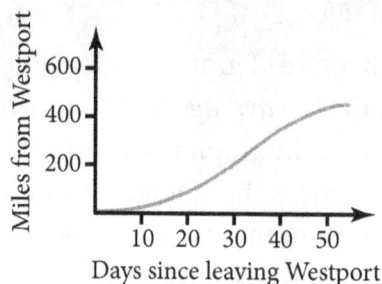

a. About how many miles per day did the wagon train travel for the first 20 days? Explain how you got your answer.

b. What was the fastest pace the wagon train achieved? Explain how you got your answer.

2. Tyler Dunkalot and his elementary school friends want to get basketball uniforms. They are saving money so they can contribute to the purchase. One Friday afternoon, right after getting their weekly allowances, they each put some of the allowance into their piggy banks and count the savings so far.

Here's how much Tyler and his friends have that Friday (including the amount they just added) and how much each will add to savings every Friday from then on.

• Tyler has $2 now and will add 50¢ each week.

• Robin has $2 now and will add 70¢ each week.

• Max has $4 now and will add 70¢ each week.

• Daniel has $4 now and will add 30¢ each week.

Draw graphs showing accumulated savings versus time elapsed for Tyler and each of his friends for a 10-week period. Assume they add their savings to their piggy banks each Friday and do not spend any of the saved money.

Draw all four graphs on the same set of axes, using a different color to represent each friend.

continued ▶

3. Compare your four graphs. Describe how they are the same and how they are different.

4. For each friend, develop a formula or an equation that shows how much money is in the piggy bank *n* weeks after they first counted their savings.

What a Mess!

Part I: The Mess

An oil tanker has suffered an explosion out at sea. Thousands of gallons of oil are spreading across the ocean. Lindsay, who is flying over in her airplane, sees that the oil slick appears to be in the shape of a circle.

When Lindsay first sees the oil slick, the radius of the circle is 70 meters. She flies overhead for a while and estimates that the radius is increasing at a rate of 6 meters per hour.

1. Make an In-Out table in which the *In* is the number of hours since Lindsay first observed the oil slick and the *Out* is the radius of the oil slick after that many hours.

2. Draw a graph based on your In-Out table.

3. Find a rule for your graph and table.

Part II: The Cleanup

Seven hours after Lindsay first sees the oil spill with a radius of 70 meters, a cleanup operation begins. Rescue workers pump a special detergent into the center of the spill. As the detergent spreads, a circle of clean water develops inside the oil spill. The radius of this circle increases at 10 meters per hour.

4. What was the radius of the oil spill when the cleanup operation began? Did the rescue team start soon enough and pump fast enough to eventually counteract the oil spill? If not, why not? If so, how many hours will it take until the spill is neutralized?

Traveling Time

Abida has packed her bags and is ready to leave for college. She needs to catch an early train because she has a long way to travel.

Abida lives in Cincinnati, Ohio, and will be going to school in Philadelphia, Pennsylvania, 550 miles away. She is hoping to reach Philadelphia in time to see some historical landmarks. She figures that if she gets to Philadelphia by 3:00 p.m., she will have enough time to visit the Liberty Bell Center and Independence Hall before they close for the evening.

1. Abida's train leaves at 5:00 a.m. Assume the train ride is exactly 550 miles. How many miles per hour must the train average for her to get to Philadelphia by 3:00 p.m.? (Cincinnati and Philadelphia are in the same time zone.)

2. The 5:00 a.m. train averages 40 miles per hour for the first hour and a half. What speed must it average for the rest of the trip for Abida to reach Philadelphia by 3:00 p.m.?

3. Suppose the train actually averaged 50 miles per hour for the whole trip (which means the trip took 11 hours altogether). That doesn't necessarily mean the train traveled at a constant rate of 50 mph.

   Make up a scenario in which the train averaged 50 mph for the trip but traveled at least two different speeds along the way. Be specific about speeds, times, and distances.

continued ▶

4. Abida's train left on time at 5:00 a.m. This graph shows one possibility for how far the train traveled as a function of the time elapsed.

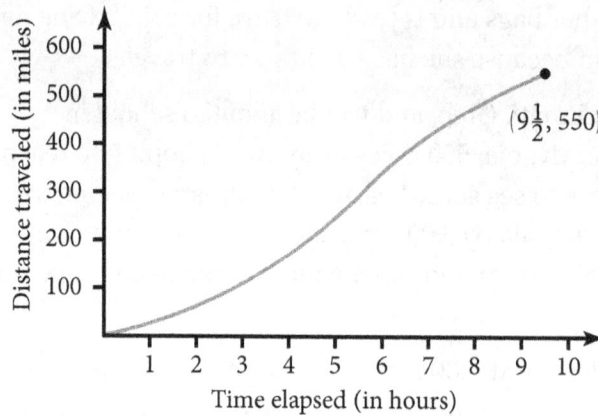

Use the graph to answer these questions

a. Was the train's average speed faster during the first 2 hours of the trip or the last 2 hours?

b. During what 1-hour period was the train's average speed the fastest?

c. If you had to pick the precise instant the train was going the fastest, what point would you choose and why?

Comparative Growth

Josh is doing a study comparing the population growth of his hometown in the early twentieth century with its population growth in the early 1980s.

When he comes across the two graphs shown here, Josh notices that the first graph appears to be steeper than the second. He concludes that the population was growing faster at the beginning of the century than in the early 1980s.

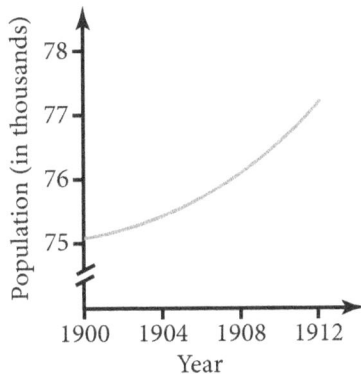

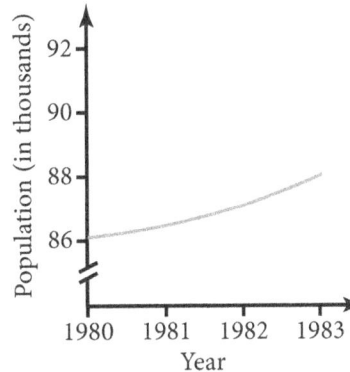

1. Use the first graph to calculate the average increase in population per year for the interval from 1900 to 1912. The graph shows the coordinates of the starting and ending points of this period.

2. Use the second graph to calculate the average increase in population per year for the interval from 1980 to 1983. The graph shows the coordinates of the starting and ending points of this period.

3. Do your results support Josh's conclusion? What does this suggest about using graphs to compare rates of change? If Josh's conclusion is incorrect, what should he do instead?

If Looks Don't Matter, What Does?

In the activity *Comparative Growth,* you saw that you can't necessarily compare rates of growth in two graphs solely from the appearance of the graphs. Now you will examine some specific situations to decide what really matters.

1. The first graph shows only two of the data points from the activity *A Crowded Place.*

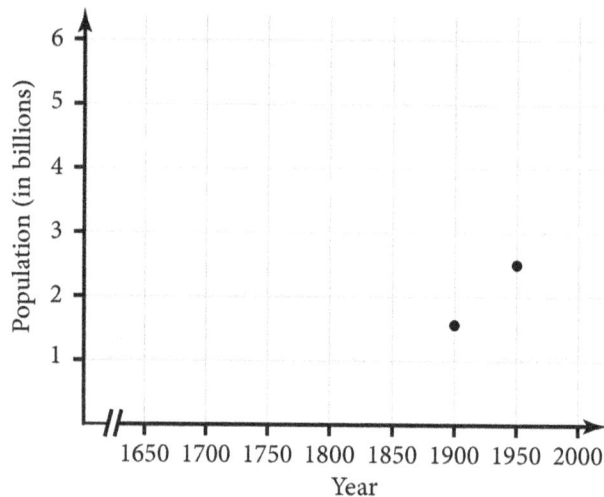

a. Use the two points to find the average population growth per year from 1900 to 1950. Explain the process you use.

b. What would this portion of the graph look like if the population had grown by the same number of people each year?

2. The graph at the right shows two points from the graph in Question 1 of *Story Sketches.* Use the graph to find the average distance traveled per day from Day 20 to Day 30. Explain the process you use.

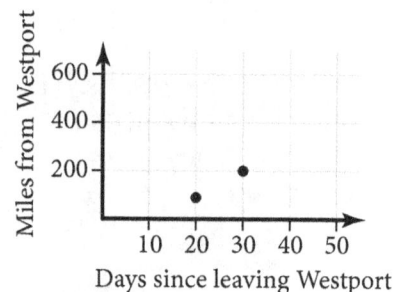

continued ▶

3. Lani is a bit older than Tyler Dunkalot and his friends. The graph below shows the amount of money Lani has in her savings account at various times over the course of several weeks.

At what rate did the amount in Lani's savings account change per week? Explain how you found your answer from the graph.

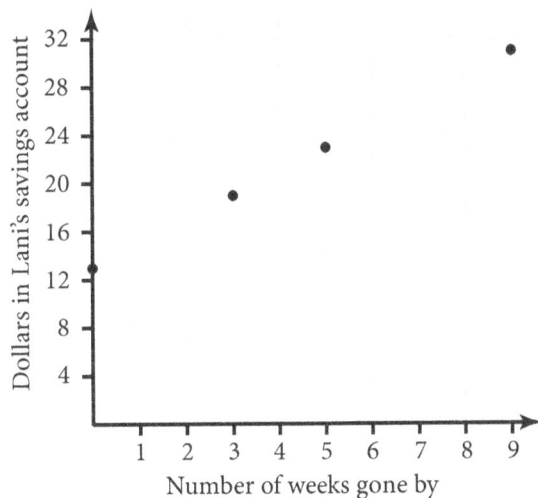

4. Find the rate of change of the function whose graph is shown below. Explain your work.

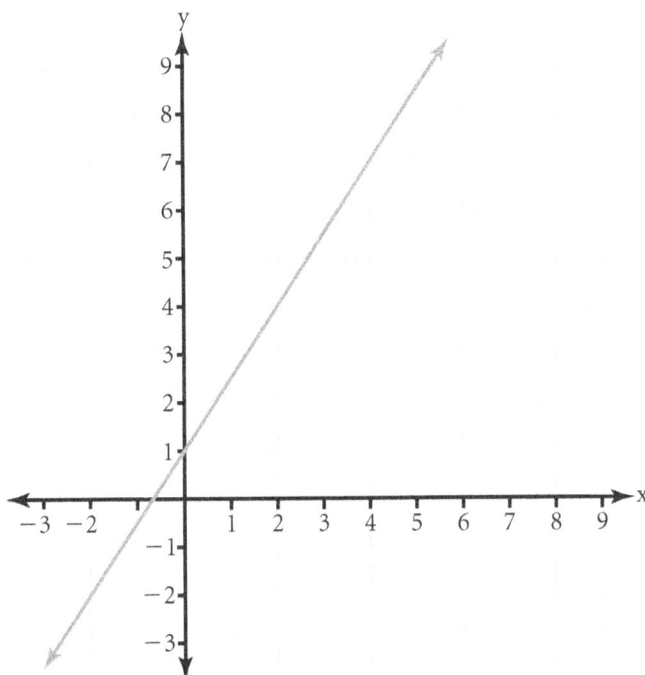

All in a Row

The simplest type of growth to study is constant growth, and that leads to work with linear functions. In the next several activities, you will develop the important concept of slope. Slope is the primary tool for describing the rate of change for a linear function.

You'll use this concept to develop equations for straight lines. You'll also see whether a linear model for population growth can accurately predict future trends.

Niall McNamara and Melyssa Brixner explore connections between graphs and formulas.

Formulating the Rate

In each situation in *If Looks Don't Matter, What Does?*, you used coordinates of points on a graph to find a rate of change. You were given specific numbers for the coordinates, or you could at least estimate the values of the coordinates from the graph.

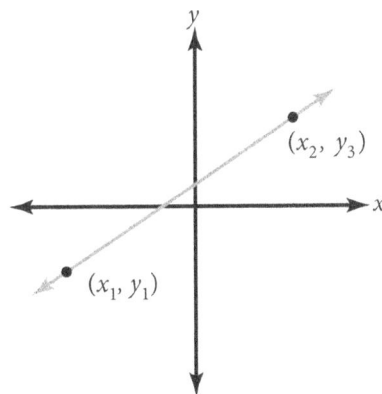

Now suppose you have a function whose graph is a straight line and that (x_1, y_1) and (x_2, y_2) are two points on the graph. That is, instead of numeric values, variables represent each of the coordinates.

Find an expression for the rate of change of the function in terms of x_1, y_1, x_2, and y_2.

Rates, Graphs, Slopes, and Equations

The concept of **slope** comes from the idea of a constant rate of change. Formally, the slope of the line connecting two points (x_1, y_1) and (x_2, y_2) is defined as the ratio

$$\frac{y_2 - y_1}{x_2 - x_1}$$

The numerator of this fraction is sometimes referred to as the *change in y* or the *rise*. The denominator is called the *change in x* or the *run*.

In this activity, you will work with rates and slopes, exploring their connections to graphs and formulas.

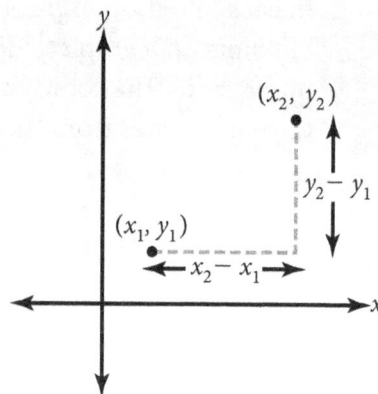

1. A jogger is moving at a rate of 500 feet per minute. That's about 6 miles per hour.

   a. Plot a graph showing the distance the jogger has traveled (in feet) as a function of t (in minutes). Use $t = 0$ as the time when the jogger starts.

   b. Find an equation for this function.

   c. Choose two points on your graph and use their coordinates to compute the slope of the graph.

2. A reservoir that is partly filled contains 200,000 cubic feet of water. Water is then added to the reservoir at a rate of 7000 cubic feet per hour.

   a. Plot a graph showing the amount of water in the reservoir as a function of t. Use $t = 0$ as the time when the water begins to be added. You may want to "skip" part of the vertical axis as was done in the graph in *Growing Up*.

   b. Find an equation for this function.

   c. Choose two points on your graph and use their coordinates to compute the slope of the graph.

continued ▶

3. a. Plot the data in this In-Out table.

In	Out
3	18
5	26
10	46
16	70

b. Choose two points on your graph and use their coordinates to compute the slope of the line connecting them.

c. Find an equation to describe your graph.

4. a. Choose two points on the line shown here and use their coordinates to compute the slope of the line.

b. Find an equation for the line.

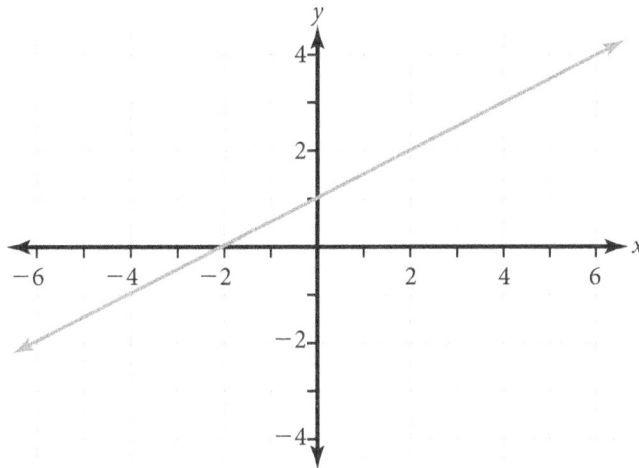

More About Tyler's Friends

In *Story Sketches,* you met Tyler Dunkalot and his friends, who were saving money to help buy basketball uniforms.

You may recall that two of Tyler's friends, Robin and Max, both saved 70¢ per week. At the start of their savings program, Robin had $2 and Max had $4.

1. Graph both Robin's and Max's savings as functions of the time elapsed. Treat these as *continuous* functions, as if Robin and Max were each saving their allowances gradually throughout the week at a constant rate. Use $t = 0$ to represent the moment when they started saving, and use the same set of axes for both graphs.

2. Find the slope of each of your graphs. Explain what you notice about the slopes.

3. Will the two graphs ever intersect? Explain your answer.

4. a. Find a formula for the amount of money each of the friends has at time t.

   b. How are these formulas related to your answers to Question 2?

Wake Up!

Getting going in the morning was probably hard for folks on the Overland Trail (just as it may be for you today), so coffee was a highly valued commodity.

The Cazneau family starts their journey with a supply of 30 pounds of coffee. When they arrive at Sutter's Fort in California, 188 days later, they have only 3 pounds left.

1. Draw a graph showing the amount of coffee remaining as a function of the time elapsed since the start of the journey. For simplicity, assume the Cazneaus consume coffee at a constant rate.

2. Find the average amount of coffee consumed per day.

3. Develop an equation for the function represented by your graph.

4. Find the slope of your graph.

5. Suppose that at the beginning of their 188-day journey, the Cazneaus are 1600 miles from Sutter's Fort.

   a. How many miles per day do they travel? For simplicity, assume they travel the same distance each day.

   b. Of course, as the Cazneaus travel, their distance from Sutter's Fort decreases. Write a formula expressing the distance remaining as a function of the number of days they have been traveling.

   c. Make a graph of your function from part b.

   d. Find the slope of your graph.

California, Here I Come!

According to the U.S. census,
the population of California
in 1850 was about 92,600.
During the 1850s, with the
great westward migration, the
population grew substantially.
In 1860, the census population
was 380,000.

1. What was the average annual
   population increase during
   the 1850s (that is, from 1850
   to 1860)?

2. Suppose California's population had continued to increase
   after 1860 by the same annual amount as it averaged during
   the 1850s. What would the population have been in each of
   these years?

   a. 1900

   b. 1950

   c. 2000

3. Generalize your results from Question 2. That is, develop
   a formula for California's population in year X. Base your formula
   on the assumption that the population continued to increase
   after 1860 by the same annual amount as it averaged during
   the 1850s.

4. According to the U.S. census, California's population in 2000
   was 33,871,648. How does this amount compare to the figure
   you found in Question 2c? What do you think accounts for
   the difference between your prediction and the actual population
   in 2000?

continued ▶

Historical note: At the time of the gold rush (the 1850s), the U.S. Constitution stated that taxes and representation in Congress would be apportioned among the states "according to their respective numbers, which shall be determined by adding to the whole number of free persons, including those bound to service for a term of years, and excluding Indians not taxed, three-fifths of all other persons." Thus, Native Americans were not counted at all unless they were taxed, and slaves were counted as three-fifths of a person each. The Fourteenth Amendment (1868) changed the Constitution to count everyone except "Indians not taxed."

Planning the Platforms

The Platform Display

River City is getting ready for the big Fourth of July band concert that precedes the fireworks. The concert is always a major event, but this year the bandleader, Kevin, plans to make it better than ever.

Kevin wants each of the baton twirlers to stand on an individual platform, as shown here.

The twirlers will toss batons up and down to one another. Kevin wants the difference in height from one platform to the next to be the same in each case.

Kevin's Decisions

Kevin has several decisions to make.

- He needs to decide on the number of platforms. He isn't sure how many of his baton twirlers will be good enough to perform by the Fourth of July.

- He needs to decide on the height of the first platform. This will depend on how tall the baton twirler on the first platform is, and Kevin hasn't decided who the first twirler will be.

- He needs to decide on the difference in height from one platform to the next. Kevin doesn't know yet how high the twirlers will be able to toss their batons.

continued

○ Camilla's Dilemma

Camilla is in charge of building and decorating the structure. She needs a permit from the city to build the structure, so she needs to know how high the tallest platform will be.

She plans to hang a colorful strip of material from the front of each platform. Each strip will reach from the top of the platform to the ground. The width of the material is the same as the width of each platform, so she needs only one strip per platform.

Camilla needs to know the total length of material she should buy, but she can't determine that until Kevin makes his decisions.

○ Your Task

You are Camilla's assistant. She has asked you to be ready to give her the information she needs as soon as Kevin makes his decisions.

Your task in this POW is to create two formulas that will allow you to compute the total length instantly. One formula should tell you the height of the tallest platform. The other should tell you the total length of material Camilla will need. Your formulas should give these results in terms of the number of platforms, the height of the first platform, and the difference in height between adjacent platforms.

○ Write-up

1. *Problem Statement:* State the problem mathematically, independent of the context of the platforms, batons twirlers, and so on.

2. *Process:* Give details of any specific examples you worked out. Explain how those examples helped you develop the formulas.

3. *Solution:* Give the two formulas and explain why they work.

4. *Self-assessment*

Points, Slopes, and Equations

In *California, Here I Come!*, you saw that knowing two data points allows you to develop a linear model for population growth. This makes geometric sense, because two points determine a straight line.

When you actually wrote the equation, you may have worked primarily with the slope (the rate of growth) and one of those points. This also makes geometric sense, because a line can also be determined by knowing one point and the "steepness" of the line (the slope).

The questions in this activity are similar to your work in *California, Here I Come!*, except they do not involve population. Each question in Part I gives you the slope and one point of a line. Each question in Part II gives you two points in a line. For Part II, you may want to find the slope as a first step in writing the equation. Part III examines the special cases of horizontal and vertical lines.

Part I: Point-Slope Equations

1. Find an equation for the line with slope 5 that goes through the point $(3, 2)$.

2. Find an equation for the line with slope 10 that goes through the point $(-4, 7)$.

3. Find an equation for the line with slope -3 that goes through the point $(5, -4)$.

4. Find an equation for the line with slope $\frac{2}{3}$ that goes through the point $(4, 6)$.

Part II: Two-Point Equations

5. Find an equation for the line that goes through the points $(6, 2)$ and $(8, 8)$.

6. Find an equation for the line that goes through the points $(1, 5)$ and $(3, -1)$.

7. Find an equation for the line that goes through the points $(-7, 2)$ and $(-2, 5)$.

continued ▶

Part III: Horizontal and Vertical Lines

8. Find an equation for the horizontal line that goes through the point $(5, 1)$. What is the slope of this line?

9. Find an equation for the vertical line that goes through the point $(2, -6)$. What is the slope of this line?

The Why of the Line

You've seen that the slope between any pair of points on a given line will be the same. But why is this so?

Consider this diagram, which shows four points—A, B, C, and D—on the same line. The diagram also shows right triangles with segments $\overline{AB}$ and $\overline{CD}$ as their hypotenuses. The lengths of the legs of these right triangles are represented with the variables r, s, t, and u.

Use ideas from geometry to explain why the slope for the pair of points A and B must be the same as the slope for the pair of points C and D. That is, explain why the ratio $\frac{r}{s}$ must equal the ratio $\frac{t}{u}$.

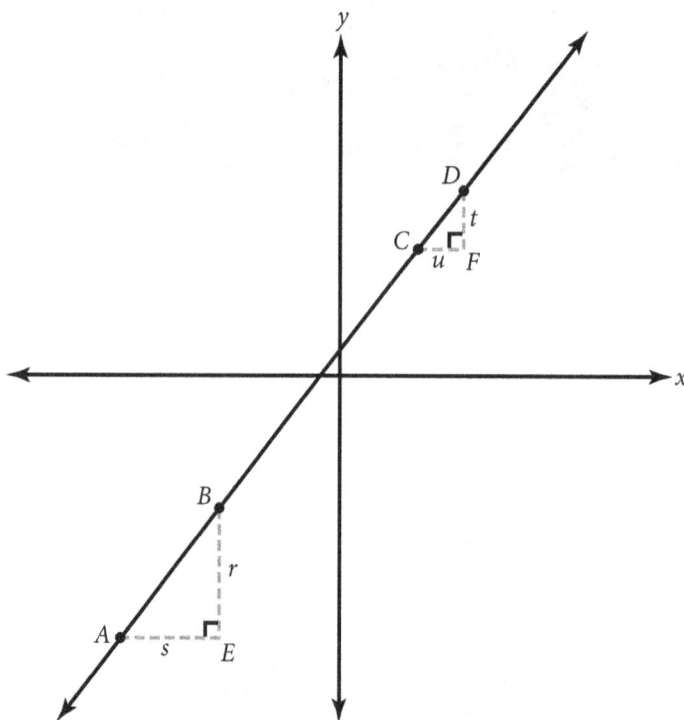

To the Rescue

A helicopter is flying to drop a supply bundle to a group of firefighters who are behind the fire lines. At the moment the helicopter crew makes the drop, the helicopter is hovering 400 feet above the ground.

The principles of physics that describe the behavior of falling objects state that when an object is falling freely, it goes faster and faster as it falls. In fact, these principles provide a specific formula describing the object's fall, which can be expressed this way.

If an object is dropped from a height of N feet, then $h(t)$, its height (in feet) off the ground t seconds after it is dropped, is given by the equation $h(t) = N - 16t^2$.

In the case of the falling supplies, the formula is $h(t) = 400 - 16t^2$, because the supply bundle is 400 feet off the ground when it starts to fall.

1. What is $h(3)$? That is, what is the height of the supply bundle 3 seconds after it is dropped?

2. How far has the supply bundle fallen during those 3 seconds?

3. What is the supply bundle's average speed during those 3 seconds?

4. How long does it take the supply bundle to reach the ground?

Beyond Linearity

Slope is an excellent concept for working with straight-line graphs. But what about rates of change that are not constant?

With the activity *The Instant of Impact,* you will begin an exploration of the meaning of an **instantaneous rate of change.** You will then refine this idea and explore it from several different perspectives.

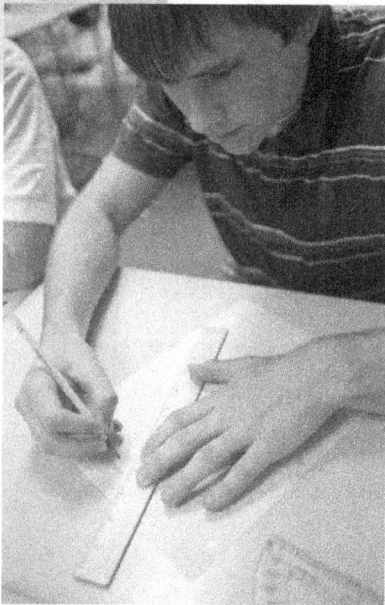

Andrew Hawn uses the slope of a tangent line to determine an instantaneous rate of change.

The Instant of Impact

In *To the Rescue*, a supply bundle is dropped from a helicopter. The bundle's height off the ground t seconds after it is dropped is given by the equation

$$h(t) = 400 - 16t^2$$

with $h(t)$ measured in feet.

Suppose the bundle can withstand the impact of hitting the ground only for speeds up to 165 feet per second. To determine whether the bundle will survive the fall, you need to find out how fast it is traveling when it hits the ground.

Keep careful track of your computations in this activity so you can compare results from question to question.

1. a. How far does the bundle fall during its last 2 seconds (that is, during the time from $t = 3$ to $t = 5$)?

   b. What is the bundle's average speed during that 2-second interval?

2. a. How far does the bundle fall during the last second before it hits the ground?

   b. What is the bundle's average speed during that last second?

3. What is the bundle's average speed during the last half-second before it hits the ground?

4. What is the bundle's average speed during the last tenth of a second before it hits the ground?

5. What is the bundle's speed at the exact moment it hits the ground?

Doctor's Orders

Clayton is a stunt diver. His most famous dive is off a cliff into the ocean near the town of Poco Loco.

Unfortunately, Clayton has injured his wrist. The impact of hitting the water at a high speed could do further damage. His doctor recommends that he do no dives in which the speed of his entry into the water is greater than 60 feet per second.

The cliff at Poco Loco is 62 feet high. Clayton always begins his dive with a jump, so he actually starts his fall from a height of 64 feet. Therefore, his height above the ocean is given by the formula

$$h(t) = 64 - 16t^2$$

where t is the time (in seconds) from when he begins his fall and $h(t)$ is his height (in feet) above the ocean.

1. How high above the ocean is Clayton 1 second after he begins his fall?

2. What is the value of t when Clayton hits the water?

3. What is Clayton's average speed during the final second of his dive?

4. What is Clayton's average speed during the final half-second of his dive?

5. Can Clayton perform his famous dive without violating his doctor's instructions? Explain.

Photo Finish

Speedy is the star runner of her country's track team. Among other events, she runs the last 400 meters of the 1600-meter relay race.

A sports analyst studied the video of one of Speedy's races. The analyst came up with this formula to describe the distance Speedy had run at a given time in the race.

$$m(t) = 0.1t^2 + 3t$$

In this formula, $m(t)$ gives the number of meters Speedy had run after t seconds, with time and distance measured from the beginning of her 400-meter segment of the race. (Although this formula might not be very accurate, treat it as if it were completely correct.)

1. How long did it take Speedy to finish the race? That is, how long did it take her to run 400 meters? Explain how you found your answer.

2. The analyst photographed Speedy at the instant she crossed the finish line. The photo is slightly blurred, so you know Speedy was going pretty fast, but you can't tell her exact speed at the instant the photo was taken.

   Find Speedy's speed at that instant.

3. a. Find Speedy's speed at three other instants during the race.

   b. Was there an instant when Speedy was going exactly 10 meters per second? If so, when was that instant?

Speed and Slope

Part I: An Instantaneous Summary

You know that average speed can be defined by the simple formula

$$\text{average speed} = \frac{\text{distance traveled}}{\text{time elapsed}}$$

You've also realized that instantaneous speed is a more complicated idea.

Summarize what you have learned about how to calculate instantaneous speed from *The Instant of Impact, Doctor's Orders,* and *Photo Finish.*

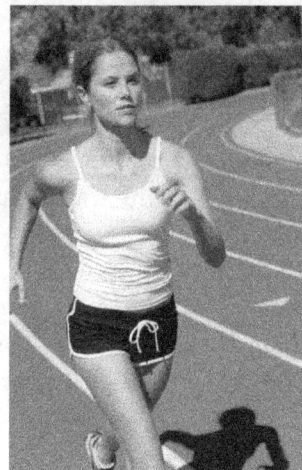

Part II: A Linear Review

You have recently investigated how rate of change might be calculated for nonlinear situations. But you shouldn't forget how to use constant rates and the slope of a straight line, whether in the context of a real-life situation or in terms of a graph. Here is one problem of each type.

1. Zoe likes to keep track of rainfall in her area by watching the level of water in a barrel outside her house. A steady rainstorm starts one morning. At one point, Zoe observes that the water level is 420 millimeters. Three hours later, the level has increased to 438 millimeters.

   Assume the rain continues steadily, so that the water level rises at the same rate throughout that 3-hour period. Use t to represent the time elapsed (in hours) from Zoe's first measurement, and write a function expressing the water level in terms of t.

2. Find an equation for the line that goes through the points $(5, -8)$ and $(13, 4)$.

ZOOOOOOOOM

In *Photo Finish,* the function $m(t) = 0.1t^2 + 3t$ describes the distance Speedy had run in terms of time elapsed. Graph this function, and adjust the viewing window so your graph includes the point (50, 400).

Because Speedy completed her 400-meter run in 50 seconds, this point is on the graph. Zoom in on the point (50, 400) until the graph on your screen looks pretty much like a straight line.

1. a. Trace the graph to find the coordinates of two points on this apparent straight line. Write down those two coordinate pairs.

   b. Find the slope of the line connecting those points.

2. What does the slope you found mean in terms of Speedy's race?

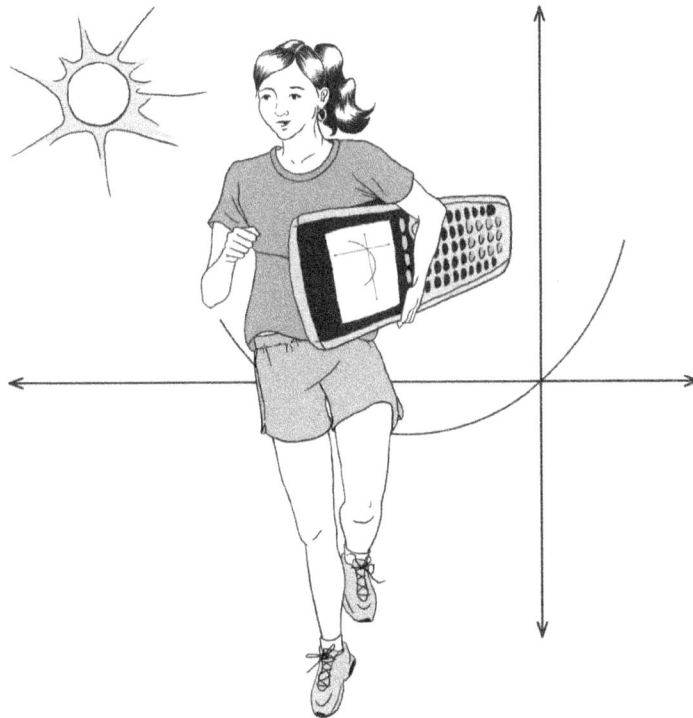

The Growth of the Oil Slick

The idea of an instantaneous rate of change applies to more than speed. This activity concerns the rate of growth of the area of the oil slick from *What a Mess!*

When Lindsay first spotted the circular oil slick, the radius of the circle was 70 meters. She noticed that the radius was increasing by 6 meters per hour. This means the radius can be described by the formula $r = 70 + 6t$, where t is the number of hours since Lindsay first saw the oil slick. So the area in square meters of the oil slick after t hours is given by the function

$$A(t) = \pi(70 + 6t)^2$$

1. What was the area covered by the oil slick when Lindsay first saw it (at $t = 0$)?

In the remaining questions, express the rate of growth of the oil slick in square meters per hour.

2. a. At what average rate did the oil slick grow during the first 2 hours after Lindsay's initial observation?

   b. At what average rate did the oil slick grow during the first half-hour after Lindsay's initial observation?

   c. At what average rate did the oil slick grow during the 15 minutes before Lindsay's initial observation?

3. At what rate was the oil slick growing at the instant when Lindsay first saw it?

Speeds, Rates, and Derivatives

The derivative of a function at a point is one of the basic concepts of calculus. If a function is defined by the equation $y = f(x)$ and (a, b) is a point on the graph, then the **derivative** of f at (a, b) can be thought of in at least two ways.

- It is the slope of the line that is tangent to the curve at (a, b).

- It is the instantaneous rate at which the y-value of the function is changing as the x-value increases through $x = a$. We often call this the derivative at $x = a$ rather than the derivative at (a, b).

In this activity, you will work with this new idea in connection with some familiar situations. Keep in mind that you can find derivatives by using smaller and smaller intervals around a particular value of a.

1. The function $h(t) = 400 - 16t^2$ gives the height of the supply bundle (in feet) t seconds after it is dropped from a height of 400 feet.

   a. Find the derivative of this function at the point $(3, 256)$.

   b. What does your answer tell you about the speed at which the supply bundle is falling?

2. The function $A(t) = \pi(70 + 6t)^2$ gives the area of the oil slick (in square meters) t hours after Lindsay first spots it.

   a. Find the rate at which the area is growing exactly one hour after Lindsay first saw it.

   b. Express your answer as a derivative.

continued ◗

3. The activity *Wake Up!* described the Cazneau family's consumption of coffee. The function $f(d) = 30 - 0.14d$ gives a good approximation of the amount of coffee (in pounds) the Cazneau family had left after d days.

For this question, treat this as a continuous function, rather than as a discrete function. That is, assume d need not be a whole number.

a. Because $f(5) = 29.3$, the graph of this function goes through the point (5, 29.3). What is the derivative of the function at this point?

b. What does your answer tell you about the rate at which the Cazneaus drank coffee?

c. Pick a point on the graph of the function other than (5, 29.3). Find the derivative at that point.

Zooming Free-for-All

You saw in *ZOOOOOOOOM* that if you zoom in on the graph of the function $m(t) = 0.1t^2 + 3t$, the graph quickly begins to look like a straight line. (Of course, you have to take into account that on a calculator screen, even the graph of a linear function might not look perfectly straight.)

Now you will investigate whether this phenomenon occurs for other functions as well.

1. Start with a graph of the function $h(t) = 400 - 16t^2$, which represents the height of the falling supply bundle. Choose a point on the graph and zoom in on that point.

   a. Does the graph begin to appear straight?

   b. If the graph does appear straight, what does the slope of that apparent straight line mean in terms of the falling supply bundle?

2. Experiment with other functions and other points. Pick a function, choose a point on its graph, and zoom in on that point. Don't worry about whether you can find a meaningful real-life situation for the function.

   a. Does the graph begin to appear straight?

   b. If the graph does appear straight, what does a straight line with that slope through your given point represent in terms of the graph?

3. Can you find a graph and a point on the graph so that, no matter how much you zoom in on that point, the graph will still not appear straight? You might try to sketch such a graph by hand, even if you can't find a formula for it.

On a Tangent

A **secant line** for the graph of a function is the line (or line segment) connecting two points on the graph. A **tangent line** is a line that "just touches" the graph at a point. In this activity, you will explore these two geometric concepts and their connections with derivatives.

1. Consider the function f defined by th e equation $f(x) = 0.5x^2$.

   a. Sketch the graph of this function, with the scale on your x-axis going from -1 to 3. Use a full-size sheet of graph paper so you will be able to get enough detail in part c.

   b. Label the point $(2, 2)$ on your graph.

   c. The points listed here are also on your graph. In each case, draw the secant line connecting the point to $(2, 2)$ and find the slope of that secant.

   i. $(0, 0)$ ii. $(1, 0.5)$ iii. $(1.5, 1.125)$ iv. $(1.9, 1.805)$

   d. Draw the line that is tangent to your graph at $(2, 2)$. Estimate the slope of that tangent line and explain your reasoning.

   e. Find the derivative of the function f at the point $(2, 2)$.

2. Consider this graph of a function.

   a. Make a copy of this graph.

   b. Draw the tangent lines to the graph at each of points A, B, and C.

   c. Use your tangent lines to estimate the derivative of the function at each of points A, B, and C.

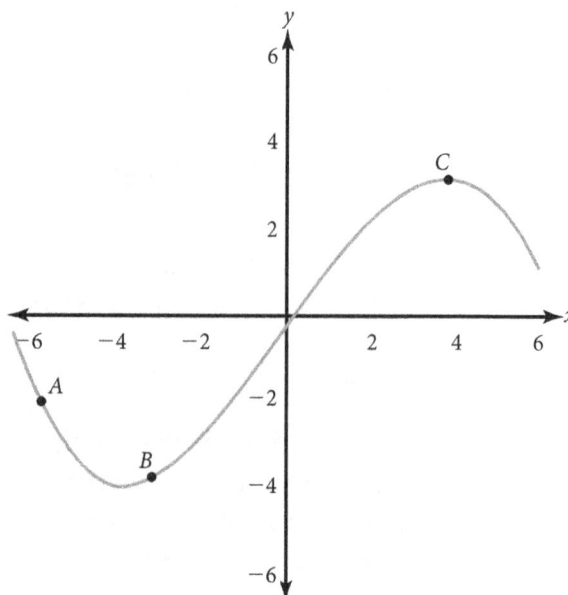

Around King Arthur's Table

King Arthur, the ruler of Camelot, loves inviting his knights over for parties around his round table.

If there is something pleasant the king can give to only one knight—like an extra dessert or a dragon to chase—he has them play a game to determine who will get it.

The game goes like this.

First, King Arthur puts numbers on the chairs, beginning with 1 and continuing around the table, with one chair for each knight. He has the knights sit down so every chair is occupied.

The king then stands behind the knight in chair 1 and says, "You're in." He moves to the knight in chair 2 and says, "You're out," and that knight leaves his seat and goes to stand at the side of the room to watch the rest of the game. The king next moves to the knight in chair 3 and says, "You're in." Then he says, "You're out" to the knight in chair 4, and that knight leaves his seat to stand at the side of the room.

The king continues around the table in this manner. When he comes back around to the knight in chair 1, he says either "You're in" or "You're out," depending on what he said to the previous knight. If the previous knight was "in," the knight in chair 1 is now "out," and vice versa.

The king keeps moving around and around the table, alternately saying, "You're in" or "You're out" to the knights who remain at the table. If a chair is now empty, he simply skips it. He continues until only one knight is left sitting at the table. That knight is the winner.

continued

Your Task

The number of knights varies from day to day, depending on who is sick and who is out chasing dragons. Sometimes there are only a few, and sometimes there are over a hundred!

Here's the big question of this POW.

If you know how many knights are going to be at the table, how can you quickly determine which chair to sit in so that you will win?

Your task is to develop a general rule, formula, or procedure that will predict the winning seat in terms of the number of knights present. Be sure to explain why your rule works.

Write-up

1. *Problem Statement*

2. *Process*

3. *Solution*

4. *Self-assessment*

What's It All About?

The idea of "the derivative of a function at a point" is important in mathematics. In fact, it plays a key role in calculus.

You will be using this idea more in the unit. Thus, it would be a good idea for you to pull together what you know about it now, so you will be ready to build on that knowledge.

Write down what you have learned so far about the idea of a derivative. Focus on these issues.

* What it means
* How you calculate it
* How it relates to other important ideas

Be sure to include specific examples.

A Model for Population Growth

You've learned that linear functions are not necessarily good for modeling population growth. So what kind of function should you use?

In the upcoming activities, you'll discover that for a certain family of functions, the derivative has a special property that makes it an excellent candidate for modeling population growth.

Erika Cohen and Alex Ryane are writing equations based upon patterns they observed from the tables they completed for Slippery Slopes.

How Much for Broken Eggs?!!?

Do you remember a POW from Year 1 called *The Broken Eggs*? It involved a farmer whose cart was hit as she was taking her eggs to market. She wasn't injured, but her eggs were broken. You spent a while figuring out how many eggs there were.

The cost of replacing 301 eggs may not have amounted to much then, but as prices rise, it could get expensive. In this activity, you'll investigate what would happen if prices went up by the same percentage each year.

Assume that at the end of 2000, a dozen eggs cost 89¢. Also assume that prices rise 5% every year. In a situation like this, the figure of 5% is called the *rate of inflation.*

In the first two questions, you will begin to analyze the situation.

1. a. How much did a dozen eggs cost at the end of 2001?

   b. How much did the price go up during 2001?

2. a. How much did a dozen eggs cost at the end of 2002?

   b. How much did the price go up during 2002?

Gather similar information for other years, until you think you understand what's happening. Then answer these questions, assuming the 5% inflation rate continues.

3. How much will a dozen eggs cost at the end of 2100? Explain your answer.

4. In what year will a dozen eggs first cost over $100?

Small but Plentiful

The activity *How Much for Broken Eggs?!!?* involves growth in prices. This activity involves a type of population growth and is somewhat like the POW *Growth of Rat Populations* in Year 2.

As you might recall, that POW was pretty complicated. You had to keep track of males and females, of different generations of rats, of when the females were ready to give birth, and so on. The situation in this activity is much simpler.

Imagine a microscopic creature like an amoeba. Suppose that whenever one of these creatures reaches a certain size, it splits into two. Then these two amoebas each grow. When they get big enough, they each split into two, making four altogether, and so on.

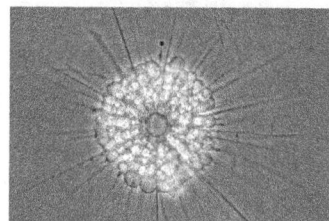

Suppose that at 12:01 a.m. on January 1 (just after midnight), there is one such tiny creature. And suppose it takes exactly 12 hours for such a creature to grow large enough to split into two, with the first split taking place at 12:01 p.m. (just after noon) on January 1.

Your task is to find a general rule for figuring out the number of creatures at a given time. (*Note:* In this model, these creatures never die—they simply split into two.)

1. Begin with specific examples by figuring out how many creatures there are at each of these times.

    a. 12:01 p.m. on January 1

    b. 12:01 a.m. on January 2

    c. 12:01 a.m. on January 5

    d. 12:01 a.m. on January 31

2. Find a general formula that tells how many creatures there are at 12:01 a.m., d days after the start of the experiment. (At 12:01 a.m. on January 1, $d = 0$.)

3. What will be the first day when there are more than one million creatures?

The Return of Alice

In the Year 2 unit *All About Alice,* Lewis Carroll's fictional character could change her height by eating special cake.

Each ounce Alice eats of a given type of cake multiplies her height by a particular factor. Alice names the different kinds of cake to match their effects. For example, if the cake multiplies her height by a factor of 3, she calls it "base 3 cake."

1. What would Alice's height be multiplied by if she eats each of these amounts of the given types of cake? Write your answers using **exponential** expressions.

   a. 4 ounces of base 3 cake

   b. 5 ounces of base 2 cake

   c. *x* ounces of base 7 cake

2. Suppose Alice eats 4 ounces of base 2 cake and then 3 more ounces of the same type of cake.

   a. Use this situation to explain the equation $2^4 \cdot 2^3 = 2^7$.

   b. Explain the equation $2^4 \cdot 2^3 = 2^7$ in terms of repeated multiplication.

3. Explain the equation $(4^6)^7 = 4^{6 \cdot 7}$ in each of these ways.

   a. In a situation involving Alice

   b. In terms of repeated multiplication

Alice realizes that she can express "cake questions" using exponential equations. For example, if she wants to know how much base 2 cake to eat to multiply her height by 32, she asks herself, "What's the solution to the equation $2^x = 32$?" She also realizes that the solution to this equation can be expressed as a logarithm: $\log_2 32$.

4. Find the value of $\log_2 32$ by solving the equation $2^x = 32$.

continued ▶

5. For each of these questions, write an exponential equation to represent the situation. Then find the numeric solution to the equation, and write the solution as a logarithm.

a. How much base 3 cake should Alice eat to multiply her height by 81?

b. How much base 2 cake should Alice eat to multiply her height by 128?

c. How much base 5 cake should Alice eat to multiply her height by 93? For this example, give your numeric solution to the nearest tenth.

Slippery Slopes

The derivatives of exponential functions show an interesting pattern. In this activity, you will find that pattern.

1. Start with the exponential function defined by the equation $y = 2^x$.

   a. Create an In-Out table like the one below. Follow these two steps to fill in several rows of your table.

      - Pick a whole-number value for x and find the y-value that goes with it.

      - Get a good approximation for the derivative of the function at the point on the graph of $y = 2^x$ that is represented by your x- and y-values.

x-value	y-value	Derivative

   b. Study the data in your table. Write an equation expressing the derivative in terms of the y-value. If necessary, add more rows of data to your table.

2. Repeat the process from Question 1, this time using the function $y = 10^x$.

3. Repeat the process from Question 1 for a third exponential function of your choice.

The Forgotten Account

Tyler and his friends (from *Story Sketches* and *More About Tyler's Friends*) had $50 left over after buying uniforms. They put the money into a bank account to give next year's team a head start.

Well, next year's team forgot about the account, and so did the following year's team. Pretty soon, the account was completely forgotten.

Suppose the account earned 4.5% interest at the end of each year. (For simplicity, assume the account was opened on January 1.) The interest is added to the amount already in the account, so each year's interest is on a larger amount than the year before. (This is called *compound interest*. More specifically, the interest for this account is said to be *compounded annually* because interest is added to the account at the end of each year.)

1. How much money was in the account 5 years after Tyler and his friends started it?

2. Find a formula that describes the amount of money in the account after t years.

3. Write an equation that could be used to figure out how many years it would take for the account to grow to $500.

4. Solve your equation.

5. Express the solution to your equation as a logarithm.

How Does It Grow?

To solve the central unit problem, you will need to find a function that grows in the same way that populations grow.

In Question 1, you will look at a simple situation to get an intuitive sense of what to expect from population growth.

1. Suppose a town has a population of 5000 and the population grows by 40 people in a single year. How much growth would you expect for a similar town with a population of 10,000? Explain your reasoning.

Questions 2 to 5 present two functions as possible models to describe population as a function of time. You will begin by getting some data for each function and its derivative.

2. Consider the linear function $f(x) = 3x + 4$.

   a. Choose three values for x. Find $f(x)$ for each value.

   b. Find the derivative of the function at each of your three x-values.

3. Now consider the quadratic function $g(x) = x^2 - 9$.

   a. Choose three values for x. Find $g(x)$ for each value.

   b. Find the derivative of the function at each of your three x-values.

In Questions 4 and 5, remember that the derivative describes a rate of growth.

4. Return to the function $f(x) = 3x + 4$ from Question 2.

   a. Describe any relationships you notice between the derivative and either x or $f(x)$.

   b. Explain why the relationships you found are or are not appropriate to use in a mathematical representation of population growth. Consider whether those relationships seem to hold true for the situation in Question 1.

continued ▶

5. Now look at the function $g(x) = x^2 - 9$ from Question 3.

   a. Describe any relationships you notice between the derivative and either x or $g(x)$.

   b. Explain why the relationships you found are or are not appropriate to use in a mathematical representation of population growth. Consider whether those relationships seem to hold true for the situation in Question 1.

The Significance of a Sign

When you are graphing a function, knowing the signs of the coordinates of the points can be helpful. For instance, those signs tell you which quadrant a point is in. And if a coordinate is 0, you know that the point is on a coordinate axis.

In this activity, you'll explore similar issues concerning the sign of the derivative of a function.

1. Make a copy of the graph. Then identify where on the graph the function's derivative is positive, where the derivative is negative, and where the derivative is 0. Remember that the derivative at a point on the function's graph can be thought of as the slope of the tangent line at that point.

2. Sketch the graph of a function for which the derivative is positive for all values of x.

3. Sketch the graph of a function for which there are exactly two points where the derivative is 0.

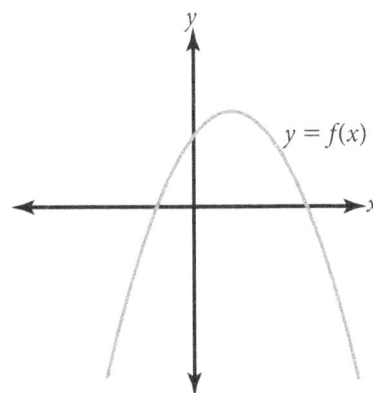

The Sound of a Logarithm

Logarithms are a convenient concept for talking about exponents. For instance, the expression $\log_{10} 564$ represents the solution to the equation $10^x = 564$. Logarithms are also used as the basis for certain units of measurement in science.

One such application concerns the measurement of noise levels. The *decibel scale* uses logarithms to describe the intensity of a sound in comparison to a particular reference value. The symbol I_0 is used to represent the intensity of a sound at the threshold of human hearing—that is, the quietest sound that humans can hear.

If I is the intensity of some other sound, the ratio of this intensity to I_0 is called the *relative intensity* of that other sound. We will represent the ratio $\frac{I}{I_0}$ by the letter R. Using this ratio, the *noise level* of that other sound, represented by N, is given by the equation

$$N = 10 \log_{10} R$$

The value of N is given in *decibels,* abbreviated as dB.

Suppose a sound has an intensity I that is 100 times the intensity of a sound at the threshold of human hearing. That is, suppose the relative intensity R is 100. Because $\log_{10} 100 = 2$, this means the sound has a noise level of $10 \cdot 2$, or 20 decibels.

A sound at the threshold of human hearing, for which I is equal to I_0, has a noise level of 0 decibels, because $R = 1$ and $10 \cdot \log_{10} 1 = 10 \cdot 0 = 0$.

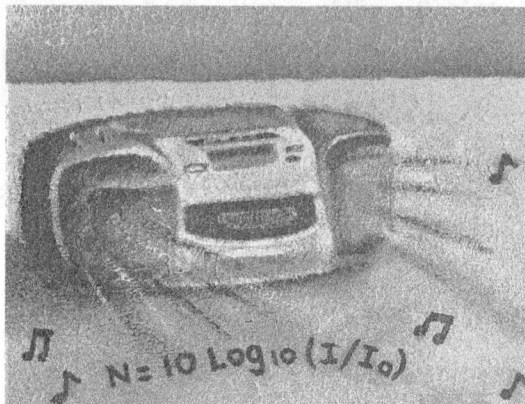

continued ▶

1. A normal conversation has a relative intensity of approximately 100,000. What is its noise level in decibels?

2. A loud police whistle has a noise level of approximately 90 decibels.

   a. How does the whistle's sound intensity compare to the threshold of human hearing?

   b. How does the whistle's sound intensity compare to ordinary conversation?

3. A sound intensity of approximately 10^{13} times the threshold level will cause a person's ears to hurt. What is the noise level of such a sound?

4. a. If one sound measures 42 decibels, and the relative intensity of a second sound is 3 times the relative intensity of the first, what is the noise level of the second sound?

   b. If one sound measures 60 decibels and another measures 68 decibels, how does the relative intensity of the second sound compare to the relative intensity of the first sound?

Adapted with permission from *College Algebra: A Preliminary Edition*, by Linda Kime and Judy Clark ©1996 by John Wiley and Sons. Reprinted by permission of John Wiley & Sons, Inc.

The Power of Powers

You saw in *Slippery Slopes* that certain functions have the special property that the derivative at each point is proportional to the *y*-value at that point.

For instance, for the function $f(x) = 2^x$, you found that an equation very much like

$$f'(x) = 0.69 \cdot f(x)$$

appears to hold true for all values of *x*. For the function $g(x) = 10^x$, the equation is approximately

$$g'(x) = 2.30 \cdot g(x)$$

The Proportionality Property

When such a relationship between the derivative and the *y*-value holds true, we say the function has the **proportionality property.** (This is not standard terminology, but is used as shorthand for a complex idea.)

When a function has the proportionality property, the ratio between the derivative and the *y*-value is called the *proportionality constant.* For instance, the numbers 0.69 and 2.30 are (approximately) the proportionality constants for the functions *f* and *g*.

Because the growth of a population is often proportional to the population itself, the proportionality property of functions like *f* and *g* makes them excellent candidates for a mathematical model of population growth.

In this activity, you will begin an exploration of which functions have the proportionality property.

continued ▸

The Examples

For each of the two functions below, go through these steps.

- Estimate the value of the derivative for at least three points on the graph.

- Based on your results, state whether you think the function has the proportionality property.

- If you think the function has the proportionality property, give an approximate value for the proportionality constant.

1. $h(x) = 2^x + 3^x$

2. $p(x) = 4.6 \cdot 2^x$

The Power of Powers, Continued

In *The Power of Powers,* you examined whether two particular functions had the proportionality property. Now you will continue your investigation of this property.

1. For each of the three functions below, go through these steps. Estimate the value of the derivative for at least three points on the graph.

   • Based on your results, state whether you think the function has the proportionality property.

   • If you think the function has the proportionality property, give an approximate value for the proportionality constant.

   a. $m(x) = 100 + 2^x$

   b. $k(x) = 0.83^x$

   c. $n(x) = 2^{3x}$

2. Based on your answers in Question 1 and your results from *The Power of Powers,* make some conjectures about the general form of functions with this special property.

The Best Base

You've learned that exponential functions seem to be an excellent choice for modeling population growth. But what base should you use? 2? 10? Does it matter?

This part of the unit begins with an activity in which you'll investigate whether bases are interchangeable. Eventually you'll discover that with regard to derivatives, one base is clearly the best choice. And you'll see that this special base has a connection with a concept that at first seems completely unrelated.

In their work with bases and exponents, Nikki Robinson and Ameilia Mitcalf discuss which might be the "best" base.

A Basis for Disguise

You've seen that the derivative of an exponential function is a fixed multiple of the function's y-value.

Another useful fact is that exponential expressions can sometimes "disguise" themselves in other number bases. For example, 81^5 can also be written as 3^{20}. Here, a power of 81 is "disguised" as a power of 3. In this activity, you will explore this change-of-base idea.

1. How can you show that 81^5 and 3^{20} are equal without finding the value of either expression?

2. Find a general rule for writing powers of 81 as powers of 3. That is, imagine that you want to put something in the box to make the equation $81^x = 3^{\square}$ true. Explain how the number that goes in the box depends on x. You might think about how much base 3 cake would have the same effect on Alice as x ounces of base 81 cake.

3. Reverse the roles of the two bases, 81 and 3, and find a general rule for writing powers of 3 as powers of 81. The Alice metaphor may help here, too.

4. Questions 1 to 3 may seem like special cases, because 81 is a whole-number power of 3.

   a. Suppose the two bases are 7 and 5. Can you find a general rule for writing 7^x as a power of 5?

   b. Examine whether there is always a general rule for writing b^x as a power of a, no matter what numbers are used for a and b. (Of course, the rule itself would have to depend on a and b.)

      Give the rule in the cases for which such a rule exists. Also describe the values for a or b for which such a rule does not exist.

Blue Book

Car dealers sometimes use the rule of thumb that a car loses about 30% of its value each year. Use this rule to answer Questions 1 and 2.

1. Suppose you bought a new car in December 2010 for $15,000.

    a. According to the rule of thumb, what would the car be worth at each of these times?

      i. December 2011

      ii. December 2015

      iii. December 2020

    b. Develop a general formula for the value of the car t years after its purchase.

2. Tara notices that a $20,000 car will lose about $6,000 of its value the first year, while a $10,000 car will lose about $3,000 of its value the first year. She reasons that because the more expensive car loses more value each year, it will eventually be worth less than the cheaper car.

    How long do you think it will take until this happens? Explain.

California and Exponents

In 1850, the population of California was about 92,600. In 1860, it was about 380,000. In *California, Here I Come!* you found a linear function that went through the points (1850, 92,600) and (1860, 380,000).

Now you will find an exponential function for the same population data and then examine its accuracy for making predictions. To make the arithmetic much simpler, treat the year 1850 as $x = 0$ and the year 1860 as $x = 10$.

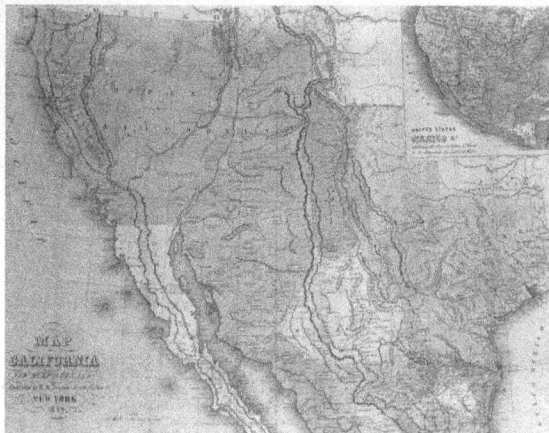

1. Find two numbers a and b so that the exponential function $y = a \cdot b^x$ goes through the points (0, 92,600) and (10, 380,000). To do this, substitute the coordinates of the first point for x and y to get an equation that involves only a. Solve that equation, and then use the second point and the value you found for a to find b.

2. Use your answer to Question 1 to determine what California's population would have been in 2000 if population growth had continued at the same growth rate as during the gold rush period.

3. Based on this exponential model, do you think the population growth rate in California has decreased or increased since the gold rush period?

Find That Base!

You know that the derivative of the exponential function $y = b^x$ is a particular constant multiplied by the y-value. As long as the base b is fixed, this proportionality constant is the same at every point on the graph. However, the proportionality constant does depend on the value of b.

Scientists prefer to use a standard base to make it easier to compare one function to another. Because any exponential function can be expressed in any base (as long as you stick to positive bases other than 1), scientists are free to pick any number for this standard base.

Because scientific work often involves derivatives, scientists have chosen for this standard base the number that makes the proportionality constant equal to 1.

Your job is to estimate this special base. In other words, estimate the value of b for which, at every point on the graph of the function $f(x) = b^x$, the derivative is equal to b^x.

Double Trouble

In inflation situations like that of *How Much for Broken Eggs?!!?*, the length of time it takes for the price to double is called the *doubling time*.

Of course, the doubling time depends on the rate of inflation. In this activity, you will investigate the relationship between these two values.

1. Consider the case of an inflation rate of 5% per year, and find the related doubling time. For simplicity, start with a price of $1 and see how long it takes until the price gets to $2. Give your answer to the nearest hundredth of a year.

2. Choose a different inflation rate, and find the related doubling time.

3. Find the doubling times for several other inflation rates. Put your results in an In-Out table in which the *In* is the inflation rate and the *Out* is the doubling time.

4. a. Look for a pattern or rule that describes your In-Out table.

   b. Try to explain why your pattern works.

5. a. For each inflation rate in your table, also compute the *quadrupling time*—the length of time it takes for the price to be multiplied by a factor of 4. If you started with $1, this would be the time required to reach $4.

   b. How does the quadrupling time compare with the doubling time? Explain your answer.

The Generous Banker

"Double Your Money in 20 Years!" read the bank's advertisement. Adam thinks this sounds like a pretty good deal, but he also thinks he might be able to talk his way into something even better. So he goes in to speak with the banker.

"Doubling your money is like increasing it by 100%, I think," he said with a practiced uncertainty. "But what if I need my money before 20 years? Can I get a proportional part of the interest each year, just in case?"

Never having studied much mathematics, the banker hesitates. "Well, that seems fair, Mr. Smith," she finally replies. "How much should you get each year?"

Adam takes out his calculator. "100% for 20 years . . . let's see . . . 100 divided by 20 . . . I guess that's 5% each year. Can you increase my account by 5% each year instead?"

The banker agrees, and Adam deposits $1,000 in an account.

1. How much will be in the account 20 years later?

2. How much will be in the account if Adam instead convinces the bank to give him a proportional amount of interest compounded every 6 months instead of once a year?

3. If Adam gets a proportional amount of interest compounded every 3 months, then how much will be in the account 20 years later?

Comparing Derivatives

Part I: Shared Points

Here are the equations for three functions whose graphs all pass through the point $(0, 0)$. The three graphs also all passthrough $(1, 1)$.

$$f(x) = x$$
$$g(x) = x^2$$
$$h(x) = x^3$$

Do you think all three functions will have the same derivative at $(0, 0)$? What about at $(1, 1)$? These questions will help you decide.

1. Draw graphs of all three functions on the same set of axes for x-values from -1 to 2. Plot enough points for each function (including noninteger values of x) to get accurate graphs. Take particular care in plotting x-values between 0 and 1.

2. Based on your graphs, answer each of these questions and explain your reasoning.

   a. Which of the three functions has the greatest derivative at the point $(0, 0)$?

   b. Which of the three functions has the greatest derivative at the point $(1, 1)$?

3. Find the actual derivative of each function at each of the two common points. Compare your results with your answers to Question 2.

continued ▶

Part II: Derivative Sketch

Suppose this is the graph of the function defined by the equation $y = f(x)$.

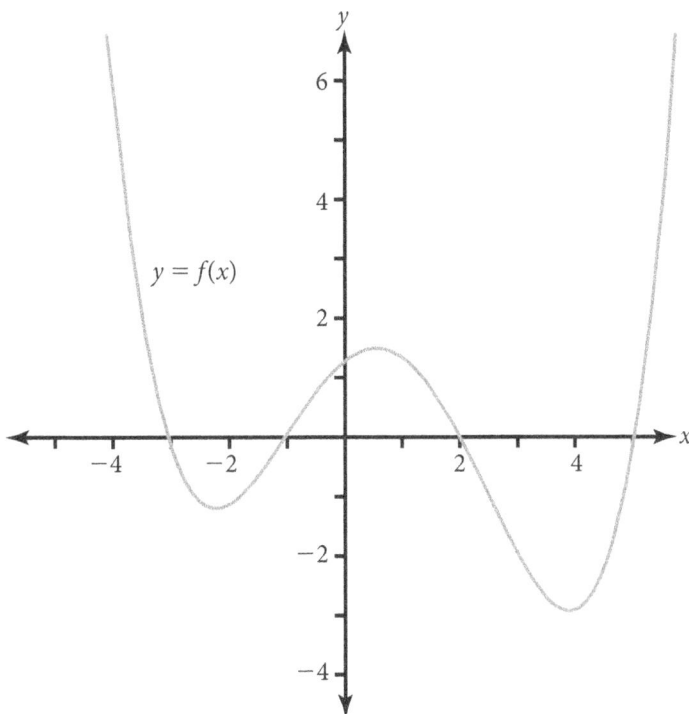

4. Make a copy of the graph. Show where on the graph the derivative of this function is positive, where the derivative is negative, and where the derivative is 0. Make your answers as complete as possible.

5. Sketch the graph of f', which is the derivative function of f. Use the same scale for the x-axis as you used for the graph of f.

The Limit of Their Generosity

In *The Generous Banker,* Adam deposits $1,000 in a bank. The bank advertised that it would double his money in 20 years. But Adam persuades the banker to increase his deposit by 5% for each of the 20 years instead of increasing it by 100% all at once at the end of the 20-year term.

Thinking he might be able to get rich because of the banker's generosity, Adam talks to the banker some more and persuades her to give him an appropriate fraction of the interest every 6 months. Then he persuades her to calculate the interest every month.

Once he has gotten this far, Adam is ready to spring his ultimate request. He asks the banker to give him proportional interest *every day* and add it to his account.

1. Without doing any calculations, about how much money do you think will be in Adam's account in 20 years?

2. Use your calculator to compute Adam's wealth after 20 years. Figure 365.25 days per year.

3. What will Adam end up with if the bank gives him proportional interest *every hour*?

California Population with e's

In the activity *California and Exponents,* you used a function of the form $y = a \cdot b^x$ to get a model of California's population growth between 1850 and 1860.

Using $x = 0$ to represent 1850 and $x = 10$ to represent 1860, you found the values of a and b so that the function $y = a \cdot b^x$ went through the points (0, 92,600) and (10, 380,000).

This activity is similar, but now you will use e as the base of the exponential function.

1. Find values for k and c so that the exponential function $y = k \cdot e^{cx}$ goes through those two points. As in the previous activity, use the first point to get an equation involving only one of the two unknown values.

2. Express the value you get for c in terms of a **natural logarithm.** A natural logarithm is a logarithm that uses the base e.

3. How are the coefficients k and c in this activity related to the coefficients a and b in *California and Exponents*? If necessary, redo Question 1 of that activity to find a and b.

Back to the Data

You're now ready to reexamine the population data from the beginning of this unit. As a first step, you'll explore how to "tweak" a function—how to make little changes in the function so its graph is closer to what you want.

You'll conclude the unit by looking for a function that comes close to the initial data, and by using that function to predict when we will all be squashed up against one another.

Kimberly Lao uses a function to predict population growth.

Tweaking the Function

Suppose you have some data points, and you need to fit them with a function. Also suppose you know more or less what type of function should fit the data and that your first guess comes pretty close to fitting, but not as close as you would like.

You then need to change the function a bit—to adjust it somehow to make it fit the data better.

How do you adjust the function to better fit the data? This activity will help you answer that question.

Begin with the function $y = e^x$. Change it in different ways and watch how the graph is affected. For instance, you might try multiplying e^x by various coefficients, or you might adjust the exponent in some way.

As you explore how the graph changes, here are some things to watch for.

- What makes the graph "more curvy" or "less curvy"?
- What changes the horizontal or vertical position of the graph?

Begin with these questions, and then investigate some others. Find out whatever you can and keep track of what you learn.

Beginning Portfolios—Part I

Think about the examples of growth and change you have studied in this unit. Focus on two particular kinds of functions.

• Linear functions, which have the form $y = a + bx$

• Exponential functions, which have the form $y = k \cdot e^{cx}$

Compare the two types of functions, addressing these issues.

• How does each function represent rates of growth or change?

• How does each function represent starting points or initial values?

• What kinds of situations is each function appropriate for describing or modeling? Be more specific than simply saying "linear growth" or "exponential growth."

Return to *A Crowded Place*

Here are the data, from the beginning of the unit, of the estimated world population over the past several centuries.

Year	Estimated Population
1650	470,000,000
1750	629,000,000
1850	1,030,000,000
1900	1,550,000,000
1950	2,560,000,000
1960	3,040,000,000
1970	3,710,000,000
1980	4,450,000,000
1990	5,280,000,000
2000	6,090,000,000
2005	6,480,000,000

As initially stated, your task is to determine, based on these data, how long it will take until people are "squashed up against one another." In this unit, that phrase is interpreted to mean that each person has exactly 1 square foot to call her or his own. Based on estimates of the earth's surface area, this means the population would have to reach approximately $1.6 \cdot 10^{15}$ people.

Here are your tasks in this final activity.

1. Plot the population data.

2. a. Find a function that approximates the data. Begin by looking for a function of the form $y = k \cdot e^{cx}$. This first approximation need not be very accurate.

continued ▸

b. Assuming the population grows according to your function, determine when each person will have only 1 square foot to call his or her own. That is, based on your function, how long will it take for the population to reach $1.6 \cdot 10^{15}$ people?

3. Repeat the process in Question 2 as often as you think is useful, looking for a better approximation. Use ideas from the activity *Tweaking the Function* to try to make whatever adjustments you think are needed.

State your best approximating function and its estimate for how long it will take the population to reach $1.6 \cdot 10^{15}$ people.

4. Discuss whether you think the population data in the table are really exponential and whether you think the data will be exponential in the future. Give reasons for your conclusions.

Beginning Portfolios—Part II

1. The beginning of this unit concentrated on linear functions. Summarize what you have learned about finding equations of straight lines, and select one or two activities that were particularly helpful to you in this area.

2. Much of this unit focused on exponential functions, a topic that leads naturally into working with compound interest.

   Pick an activity from the unit that helped develop your understanding of compound interest, and describe what you learned from the activity.

Annie Tam, Ryan Tran, and Carlos Catly spend time reviewing activities for their portfolios.

Small World, Isn't It? Portfolio

You will now put together your portfolio for *Small World, Isn't It?*
This process has three steps.

• Write a cover letter that summarizes the unit.
• Choose papers to include from your work in the unit.
• Discuss your personal growth during the unit.

Cover Letter for *Small World, Isn't It?*

Look back over *Small World, Isn't It?* and describe the central problem
of the unit and the key mathematical ideas. Your description should
give an overview of how the key ideas were developed and how they
were used to solve the central problem.

In compiling your portfolio, you will select some activities you think
were important in developing the unit's key ideas. Your cover letter
should include an explanation of why you selected each item.

Selecting Papers from *Small World, Isn't It?*

Your portfolio for *Small World, Isn't It?*
should contain these items.

• *What's It All About?*
• *Beginning Portfolios—Part I*
• *Beginning Portfolios—Part II*
  Include the activities from the unit you
  selected as part of this assignment.
• A Problem of the Week
  Select any one of the three POWs you
  completed in this unit: *The More the
  Merrier?, Planning the Platforms,* or
  Around King Arthur's Table.

continued

Personal Growth

Small World, Isn't It? focused on various kinds of growth. Your cover letter describes how the mathematical ideas developed in the unit. In addition, write about your own personal development during this unit. You may want to specifically address this question.

What changes or growth have you noticed during Year 3 in your ability to work well in groups?

Include any thoughts about your experiences that you wish to share with a reader of your portfolio.

SUPPLEMENTAL ACTIVITIES

The supplemental activities for *Small World, Isn't It?* continue the unit's areas of emphasis—rates of change and linear and exponential functions—along with a few other topics. Here are some examples.

- *Solving for Slope* and *The Slope's the Thing* give additional perspectives on how to find the slope for a linear equation.

- *Summing the Sequences*—Parts I and II are follow-ups to the POW *Planning the Platforms*.

- *Deriving Derivatives* looks at developing general formulas for derivatives.

- *Dr. Doubleday's Base* and *Investigating Constants* examine the role of the parameters in two different forms of the general exponential function.

Solving for Slope

You've seen that in an equation like $y = 3x + 5$, the coefficient of x (in this case, 3) is the slope of the graph. Now you will examine linear equations that do not have this form.

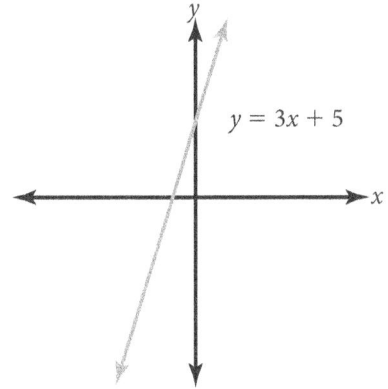

$y = 3x + 5$

1. Begin by examining these specific examples. In each case, solve the equation for y in terms of x to determine the slope of the graph.

    a. $3y - 2x = 12$

    b. $x + 4y = -8$

    c. $5x - 2y = 10$

    d. $-6y + 7x = -2$

2. Now consider the general linear equation in the standard form.

$$ax + by = c$$

    a. Develop an expression for the slope in terms of a, b, and c.

    b. Based on examples, describe how to answer each of these questions simply by looking at a, b, and c.

    • Is the line rising or falling as it goes to the right?

    • For lines rising to the right, is the line steeper than $y = x$?

    • For lines falling to the right, is the line steeper than $y = -x$? Explain not only how your method works, but also why it works.

Slope and Slant

One way to describe the "steepness" or "slant" of a line is to measure the angle formed between that line and a horizontal line. This angle is called the *angle of inclination* of the line.

For instance, this diagram shows a line *l* and a horizontal line *h*. The angle labeled θ (the Greek letter *theta*) is the angle of inclination of line *l*.

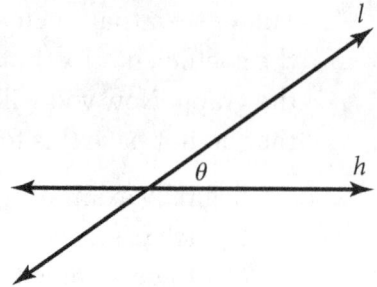

You've been working with lines that are the graphs of linear equations. In the context of the coordinate system, the angle of inclination of a line is the angle formed between that line and the positive direction of the *x*-axis. You might expect to find some relationship between the slope of such a line and its angle of inclination.

Finding such a relationship is complicated by the fact that the scales on the axes affect the steepness of the graph. In this activity, you will eliminate that complication by assuming that the vertical and horizontal axes have the same scale.

1. This diagram shows the graph of the equation $y = 2x - 1$ and two points, $(-1, -3)$ and $(3, 5)$, that fit this equation. The diagram also shows an angle of inclination θ for the line and a right triangle with $(-1, -3)$ and $(3, 5)$ as two of its vertices.

   a. Find the slope of the line $y = 2x - 1$.

   b. Use trigonometry and the right triangle to find a relationship between the slope of this line and its angle of inclination.

   c. Use your work from part b to find the measure of angle θ.

   d. Carefully draw the line through $(-1, -3)$ and $(3, 5)$ on graph paper, using the same scale for the vertical and horizontal axes. Then measure the angle of inclination and check whether it matches your result from part c.

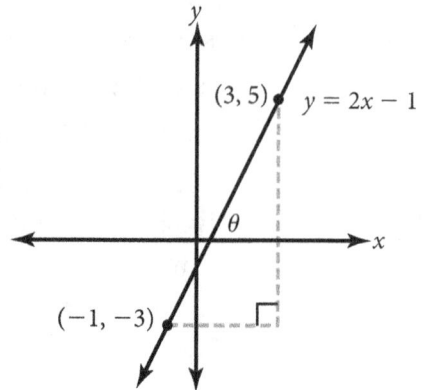

continued ▶

2. Use the ideas from Question 1 to answer these questions. Again, assume the vertical and horizontal axes have the same scale.

   a. What is the angle of inclination for a line with a slope of 1?

   b. What is the slope of a line with an angle of inclination of 30 degrees?

3. State a general principle, using trigonometry, relating the slope of a line to its angle of inclination. (*Note:* So far, we have defined the trigonometric functions only for acute angles in right triangles. Therefore, assume the line has an angle of inclination between 0 and 90 degrees.)

Predicting Parallels

When a linear equation is written to express y in terms of x, the coefficient of x is equal to the slope of the graph. For example, the graph of $y = 3x + 5$ has slope 3. This makes it easy to recognize when two distinct (that is, not equivalent) equations have graphs that are parallel, because they have the same coefficient for x.

According to this principle, the graphs of $y = 3x + 5$ and $y = 3x + 9$ should be parallel. That is, they should not have any points in common. The first task in this activity is to prove this fact.

1. Show that the equations $y = 3x + 5$ and $y = 3x + 9$ cannot have any common solutions. That is, show there is no point that is on the graphs of both equations.

What about lines that are not in "$y =$" form? Is there a simple way to recognize that two lines in the standard form $ax + by = c$ are parallel? Answer Questions 2, 3, and 4, and then try to generalize your results in Question 5.

2. a. Draw the graph of the equation $4x - 3y = 24$.

   b. Choose a point *not* on the graph. Draw a line through that point that is parallel to your graph.

   c. Find the equation of the line you drew in part b.

3. Graph the equation $8x - 6y = 24$, and compare it with the graphs from Question 2.

4. Graph the equation $20x - 15y = 120$, and compare it with the graphs from Question 2.

5. Based on Questions 2, 3, and 4, and on other equations you might examine, what general principles can you state for determining at a glance whether two linear equations in standard form have parallel graphs?

The Slope's the Thing

You know that two points determine a straight line. That means knowing the coordinates of two points on a line is enough information to get an equation for that line. You will now explore a systematic way to get the equation from that information, based on slope.

Let's call the two given points P and Q. The method is based on these two statements.

- If a point is not on that line (for instance, point A in the diagram), then the slope of the line connecting that point to P will be *different* from the slope of the line through P and Q.

- If a point is on that line (for instance, point B), then the slope of the line connecting that point to P will be the same as the slope of the line through P and Q.

1. Suppose P has coordinates (x_1, y_1), Q has coordinates (x_2, y_2), and R is another point in the plane, with coordinates (x, y).

    a. Write an expression for the slope of the line through P and Q.

    b. Write an expression for the slope of the line through P and R.

    c. Combine your expressions to get an equation stating that these two slopes are equal.

2. Apply this method to find the equation of the line through $(1, 5)$ and $(-2, -1)$.

3. Use similar triangles to prove the two statements on which this method is based.

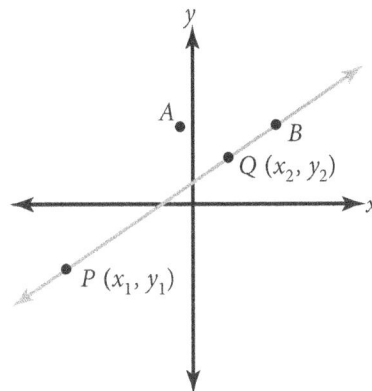

Speedy's Speed by Algebra

In the activity *Photo Finish*, Speedy the track star runs the last 400 meters of a 1600-meter relay race.

Based on a video of her race, an analyst came up with the function $m(t) = 0.1t^2 + 3t$ to describe how many meters Speedy had run after t seconds of her segment of the race. The equation $m(50) = 400$ shows that Speedy crosses the finish line after exactly 50 seconds.

In *Photo Finish*, you found Speedy's average speed for small time intervals near the end of the race. You used those average speeds to estimate her speed at the instant she crossed the finish line. You may wonder how accurate or reliable your estimate is. That is, you may be wondering this.

How fast was Speedy really going at the instant she crossed the finish line?

In this activity, you will explore an algebraic approach to answering this question.

1. First, write out in detail the computation for finding Speedy's average speed for the last tenth of a second of the race.

   a. Find out how far Speedy had run after 49.9 seconds. That is, compute $m(49.9)$.

   b. Find out how far Speedy ran during the last tenth of a second. That is, find the difference between $m(49.9)$ and 400.

   c. Find Speedy's average speed during the last tenth of a second. That is, divide the distance you found in part b by the length of the time interval, which is 0.1 second.

continued

2. Now use the process from Question 1, substituting the last h seconds for the last tenth of a second. That is, instead of working with the time interval from $t = 49.9$ to $t = 50$, work with the interval from $t = 50 - h$ to $t = 50$. Parts a to c will guide you through this process.

   a. Find out how far Speedy had run after $50 - h$ seconds. That is, find $m(50 - h)$. Your answer should be an expression in terms of h.

   b. Find out how far Speedy ran during the last h seconds. That is, write an expression (in terms of h) for the difference between $m(50 - h)$ and 400.

   c. Divide the distance you found in part b by the length of the time interval, which is h seconds.

   d. Confirm your expression in part c by substituting 0.1 for h. Does that give the same answer you found in Question 1c?

3. Your result for Question 2c should be an algebraic expression in terms of h. This expression gives Speedy's average speed during the last h seconds of the race.

   a. Simplify this expression as much as you can.

   b. What happens to your simplified expression as h gets smaller and smaller?

   c. What does your result in part b mean in terms of Speedy's instantaneous speed?

4. Does your work in Questions 1 to 3 give you further confidence in your results from *Photo Finish*? Explain.

Potential Disaster

Construction workers have partially ruptured a natural gas pipeline in a research building. The pipe's protective inner membrane has been forced out through the opening, forming a sphere-shaped balloon on the outside of the pipe.

Emergency crews are confident they can patch the hole but see no way to push the balloon back into the pipe. They decide they must cut off the balloon and then quickly patch the hole.

An electrical wire nearby poses a serious danger. If the wire and balloon come in contact with each other, the gas in the balloon will ignite and the balloon will explode, causing massive damage.

Gas is flowing into the balloon so that its volume is growing at a constant rate. The balloon was discovered 30 minutes after the rupture. In those 30 minutes, the balloon had grown to a diameter of 1 foot. The balloon will come in contact with the wire if its diameter reaches 2 feet.

It is now 20 minutes since the balloon was discovered, and the building's manager is in a panic. He thinks the balloon will explode in only 10 more minutes. But the engineer in charge of the emergency crew seems rather calm.

1. a. Why do you suppose the manager thinks they have only 10 minutes left before the balloon explodes?

   b. What is wrong with the manager's reasoning?

2. How much time does the crew really have? (The volume of a sphere of diameter d is given by the expression $\frac{1}{6}\pi d^3$.)

Proving the Tangent

In the activity *On a Tangent*, you examined the graph of the function $f(x) = 0.5x^2$ near the point $(2, 2)$. You were asked to use a series of secant lines through $(2, 2)$ to estimate the slope of the tangent line at that point, which seemed to be equal to 2.

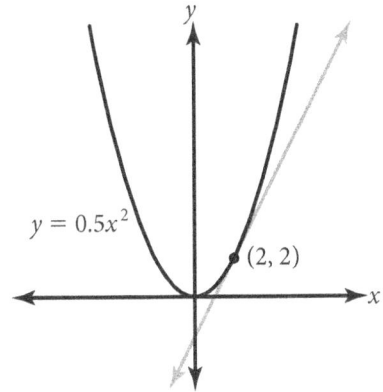

$y = 0.5x^2$

$(2, 2)$

1. Use this estimate of the derivative to write the equation of the tangent line. That is, write the equation of the line that goes through the point $(2, 2)$ and has a slope of 2.

2. Prove that your estimate of the derivative is correct by showing that $(2, 2)$ is the only point where the line you found in Question 1 meets the graph of the function $f(x) = 0.5x^2$. To do this, you will need to show that your equation from Question 1 and the equation $y = 0.5x^2$ have only one solution in common.

Summing the Sequences—Part I

In the POW *Planning the Platforms,* Camilla needs to find the sum of a sequence of numbers in which each term differs from the previous term by the same amount.

For instance, suppose the first platform is 28 inches tall, the difference in height between adjacent platforms is 8 inches, and there are 5 platforms altogether. Camilla would need to find the sum

$$28 + 36 + 44 + 52 + 60$$

A sequence such as 28, 36, 44, and so on, in which the *difference* between terms is constant, is called an **arithmetic sequence.** (In this context, the word *arithmetic* is an adjective and is pronounced ar-ith-*met*-ic, with the emphasis on the third syllable.)

The first number in an arithmetic sequence is called the *initial term.* The amount added to get each successive term is called the *difference.* In the example, the initial term is 28 and the difference is 8.

1. a. Find an expression in terms of n for the nth term of the arithmetic sequence with an initial term of 28 and a difference of 8.

   b. Find an expression in terms of n for the sum of the first n terms of this sequence.

continued ▶

If we use a to represent the initial term and d to represent the difference, then the general arithmetic sequence has the terms a, $a + d$, $a + 2d$, $a + 3d$, and so on.

2. Find an expression in terms of a, d, and n for the nth term of the general arithmetic sequence.

3. Find an expression in terms of a, d, and n for the sum of the first n terms of the general arithmetic sequence. Give your answer as an algebraic expression in closed form—that is, without using summation or ellipsis (. . .) notation.

4. Apply your expression from Question 3 to find the sum of the first 50 terms of the arithmetic sequence 15, 21, 27, 33, and so on.

Summing the Sequences—Part II

In *Summing the Sequences—Part I,* you examined arithmetic sequences, which have a constant difference between terms.

Sequences with a constant ratio between terms are called **geometric sequences.** For instance, the sequence 8, 24, 72, 216, and so on, in which each term is 3 times the previous term, is a geometric sequence. Here, 8 is the initial term and 3 is the ratio.

1. a. Find an expression in terms of n for the nth term of the geometric sequence with an initial term of 8 and a ratio of 3.

   b. Find an expression in terms of n for the sum of the first n terms of this sequence.

   To do this, compare the sum of the first n terms of this sequence with the sum of the first n terms of the sequence 24, 72, 216, and so on. The second sum is 3 times the first (because each term is 3 times the corresponding term in the original sequence). What do you get if you subtract the first sum from the second? Think about terms that cancel out. You may want to look at specific values of n.

Using a to represent the initial term and r to represent the ratio, the general geometric sequence has the terms a, ar, ar^2, ar^3, and so on.

2. Find an expression in terms of a, r, and n for the nth term of the general geometric sequence.

3. Find an expression in terms of a, r, and n for the sum of the first n terms of the general geometric sequence. Give your answer as an algebraic expression in closed form—that is, without using summation or ellipsis (. . .) notation.

4. Apply your expression from Question 3 to find the sum of the first 20 terms of the sequence 3, 6, 12, 24, and so on.

continued ◗

5. Consider the case of the geometric sequence with $a = 1$ and $r = \frac{1}{2}$.

   a. Write an expression (in terms of n) for the sum of the first n terms of this sequence. Use your work from Question 3.

   b. Find the sum of the first 10 terms of this sequence by actually adding the terms. Use it to verify your expression from part a.

   c. What happens to your expression as n increases? Does that fit your intuitive idea? Explain.

6. (Challenge) How can you generalize Question 5 to the case in which the ratio r is any number between 0 and 1?

Looking at Logarithms

Logarithms are defined in terms of exponential equations. For instance, $\log_b a$ is defined as the value of x that fits the equation $b^x = a$.

Because of this relationship between logarithms and exponents, there is a principle about logarithms corresponding to every principle about exponents.

In this activity, you will develop principles for logarithms. Throughout the activity, b represents a positive number other than 1.

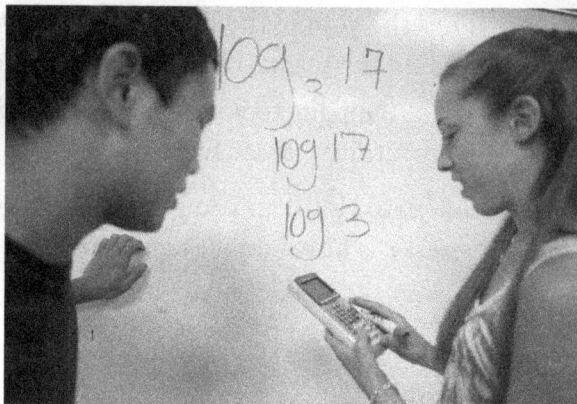

1. The additive law of exponents states

$$b^x \cdot b^y = b^{x+y}$$

Your first task is to find a corresponding principle for logarithms. Start with the specific examples in parts a and b.

   a. What is $\log_2 2^5$? What is $\log_2 2^9$? What is $\log_2 (2^5 \cdot 2^9)$?

   b. Find approximate values for $\log_3 8$, $\log_3 7$, and $\log_3 56$. What relationship do you see among these three logarithms?

   c. Generalize your results from parts a and b to get a formula for $\log_b rs$ in terms of $\log_b r$ and $\log_b s$.

   d. Use the additive law of exponents to prove your general result. (Think of r as b^x and s as b^y.)

continued ⬧

2. Another principle for exponents states

$$(b^x)^n = b^{xn}$$

Your task is to find a corresponding principle for logarithms by first looking at specific examples.

a. How does $\log_5 (5^7)^6$ compare to $\log_5 5^7$?

b. Find approximate values for $\log_3 17$ and $\log_3 17^4$. What relationship do you see between these two logarithms?

c. Generalize your results from parts a and b to get a formula for $\log_b r^n$ in terms of $\log_b r$ and n.

d. Use the principle $(b^x)^n = b^{xn}$ to prove your general result. (Again, think of r as b^x.)

3. The range, or set of possible outputs, for the function $y = b^x$ consists of all positive numbers. That is, as x varies over all possible values, y can be any positive number. What does this say about the domain, or set of possible inputs, of the function $f(u) = \log_b u$? That is, what numbers can be used for u in this function?

4. a. What is the domain of the function $y = b^x$?

b. What does your answer say about the range of the function $g(v) = \log_b v$?

Finding a Function

In the activity *The Significance of a Sign,* you were asked to sketch the graph of one function for which the derivative is positive for all values of x and another function for which the derivative is zero at exactly two points. Now you will continue that exploration.

1. Sketch the graph of a function f that fits these two conditions.
   - $f'(x)$ is positive for all values of x between -2 and 3.
   - $f'(x)$ is negative if $x < -2$ or $x > 3$.

2. Sketch the graph of a function $g(x)$ that fits these three conditions.
   - $g'(x)$ is zero for exactly three values of x.
   - $g(x)$ is positive between $x = -1$ and $x = 4$.
   - $g(x)$ is negative if $x > 4$ or $x < -1$.

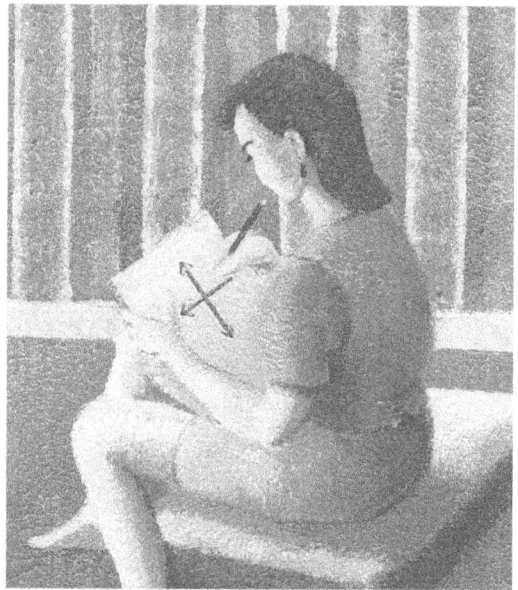

Deriving Derivatives

By looking at the derivative of a function at many of its points, you can sometimes find a rule for its derivative at any point.

In this activity, you will look for a rule that allows you to quickly find the derivative of a function of the form $y = ax^n$. The functions defined by the equations $y = 2x^2$, $y = x^3$, and $y = -3x^6$ are examples of this type of function.

Start with a specific function of this form. Make a table of values for the function's derivative, and find a rule for your table. You might then try other functions that use the same exponent and look for a generalization for functions with that exponent. This exponent, you may recall, is called the *degree* of the function.

Once you have a rule for functions of a particular degree, try a different degree and then try to generalize your results.

The Reality of Compounding

You saw in the activity *The Generous Banker* that getting 5% annual interest, compounded each year for 20 years, is not the same as doubling your money in 20 years.

1. Suppose the bank compounds interest annually. What yearly rate of interest should the bank give for money to double in 20 years?

2. Suppose the bank compounds interest quarterly. What quarterly rate of interest will double the value of an account in 20 years?

Transcendental Numbers

In the unit *Orchard Hideout*, you worked with the number π. The number π can be defined as the ratio between the circumference and the diameter of a circle. In this unit, you were introduced to the number e. Both π and e are examples of *transcendental numbers.*

What is a transcendental number? Your task is to investigate this category of numbers. Then write a report explaining what you learned about transcendental numbers and their history.

Dr. Doubleday's Base

You have seen that for any positive number b, the derivative of the exponential function $y = b^x$ is a proportionality constant multiplied by the y-value.

For example, you found that if the base b is 2, then the proportionality constant is approximately 0.69. In other words, for the function $y = 2^x$, the derivative at a point $(x, 2^x)$ on the graph is approximately $0.69 \cdot 2^x$.

You have also seen that there is a special base, called e, for which this proportionality constant is 1. In other words, for the function $y = e^x$, the derivative at any point on the graph is equal to the y-value at that point.

While it is convenient for scientists to use e as the base, Dr. Doubleday would like a base for which the proportionality constant is 2. (We don't know why.) In other words, the doctor would like to find an exponential function whose derivative at every point on its graph is equal to twice its y-value at that point.

1. Determine what base Dr. Doubleday should use. That is, find a number b such that for every point (x, b^x) on the graph of the function $f(x) = b^x$, the derivative is equal to $2 \cdot b^x$. You will only be able to estimate b. Try to get its value to the nearest hundredth.

2. Look for a relationship between your answer and the number e.

Investigating Constants

1. You've seen that the general exponential function can be written in the form $y = k \cdot e^{cx}$, where k and c are any two nonzero numbers. (What happens if k or c is zero?)

   You also know that any such function has the proportionality property. That is, the derivative at any point on the graph is proportional to the y-value at that point.

   Investigate how the value of the proportionality constant depends on the value of the parameters k and c.

2. The general exponential function can also be written in the form $y = k \cdot b^x$, where k is some nonzero number and b is a positive number other than 1. (What happens if b is equal to 1?) In this form, too, the function has the proportionality property.

   Investigate how the value of the proportionality constant depends on the value of the parameters k and b.

PHOTOGRAPHIC CREDITS

www.ingramcontent.com/pod-product-compliance
Lightning Source LLC
Chambersburg PA
CBHW051225200326
41519CB00025B/7256